Mountain Environments and Geographic Information Systems

Mountain Environments
and Geographic Information Systems

Edited by

Martin F. Price and D. Ian Heywood

Taylor & Francis
Publishers since 1798

UK	Taylor & Francis Ltd, 4 John St, London WC1N 2ET
USA	Taylor & Francis Inc., 1900 Frost Road, Suite 101, Bristol PA 19007

British Library Cataloguing in Publication Data

A catalogue record for this book is available from the British Library

ISBN 07484 0088 5

Library of Congress Cataloging in Publication Data are available

Cover design by Barking Dog Art

Typeset by Graphicraft Typesetters Ltd., Hong Kong

Printed in Great Britain by Burgess Science Press, Basingstoke on paper which has a specified pH value on final paper manufacture of not less than 7·5 and is therefore 'acid free'.

Contents

Foreword

This book undertakes the seemingly impossible task of combining the complexity of themes pertinent to mountain environments with the complexity of geographic information systems (GIS). Its preparation was catalysed by the Second International Conference on Environmental Regeneration in Headwaters, held in Sec, Czechoslovakia in November 1992 under the auspices of the World Association of Soil and Water Conservation (WASWC), the International Union of Forest Research Organizations, and many Czech organizations.

Mountain environments are characterized by their great complexity. Great variations in climate, water characteristics, soils, and geology have, over millenia, produced a wealth of ecological diversity in flora and fauna. Geomorphic processes such as surface runoff, torrential rivers, snow avalanches, or mass movements accelerate exponentially as slopes become steeper. Yet, although steep slopes and high altitudes limit and may threaten human use, well-adapted traditional land-use systems were introduced into many mountainous regions by human settlers thousands of years ago. In the past, restricted access to these regions has presented opportunities for cultural adaptation; today, it often poses a limitation to modern development.

The problems of mountain environments today, as addressed in the chapters of this book, can be characterized as follows: mountains have highly diverse natural and cultural resources, only a fraction of which can be used sustainably. There are difficulties in regional planning owing to great differentiation and to the problems of natural hazards. Because mountain resources are abused, as in the case of deforestation, there is an urgent need for the conservation of wildlife, wildlife habitats, and natural resources in general. Mountains provide relevant indicators for monitoring climate change. Finally, there is a great demand for better knowledge on which to base sustainable development under fragile conditions, and for the prediction and simulation of future changes using scenario techniques.

Like mountain environments, GIS approaches are characterized by their complexity. For geographers, the overlay of spatial, i.e. geographic, information has always been part of their profession. Whenever different maps, possibly at differing scales and with different projections, have to be compared and interpreted simultaneously, geographers have been welcomed by practitioners such as landscape architects, land-use planners, watershed managers, and decision-makers. This traditional artwork is now being greatly aided by the computer and by novel information technologies. The term Geographic Information System has become a key word almost as fascinating as the term 'sustainable development'. GIS promises easy utilization of spatial information for any purpose, be it scientific research, urban or rural planning, infrastructure development, remote sensing, or numerous other applications. That GIS is coupled with computer hardware and software makes it even more attractive to our information-oriented society.

As a consequence, there is hardly any institution charged with the handling of spatial information which is not investing in GIS. And here is where we must face reality. GIS is but a tool – an instrument. Unfortunately, it is not an instrument that is easy to use. As with all computer developments, most of the effort has been invested in the methodological and technical developments of the instrument itself. Practitioners have had relatively little input into the technology and, so far, have produced only limited practical and usable outputs. Fortunately, despite all this, a growing number of people – scientists, technicians, and planners who in the recent past have invested a great deal of effort, resources, and time to make GIS a problem-solving instrument – are now in a position to show how useful it is. This includes the authors of the chapters in the present volume, who try to marry the complexities of the mountains to the complexities of GIS options.

The World Association of Soil and Water Conservation (WASWC), an expert group of over 750 people from more than 70 countries around the globe, was associated with the Sec conference thanks to the efforts of one of its vice-presidents, Martin Haigh of Oxford Brookes University. WASWC is particularly interested in linking its principal focus, 'the propagation of the wise use of soil and water resources worldwide', with the complex theme of mountains and GIS approaches. Soil and water are precious mountain resources. Those dedicated to preserving soil and water for future generations have the conservation and sustainable use of these resources as their primary goal. Unfortunately, this goal is not shared by all users of mountain resources, nor is it perceived by many of those who indirectly affect mountain resources, such as industries that cause air pollution or governments whose policies negatively influence the development of the generally neglected mountain regions in their countries.

There is a great need for more information that can be made easily available to decision-makers. GIS can be a method that demonstrates the implications of political, economic or ecological decisions for mountain environments. In addition to GIS tools, creative models are being developed to show trends, to devise credible scenarios, to present convincing arguments that will change negative attitudes and overcome lack of perception, and to induce the stabilizing interventions that are so badly needed.

The World Association of Soil and Water Conservation will do its best to propagate the messages contained in this book. WASWC would like to extend its thanks to the editors, Martin Price and Ian Heywood, and the publisher, Taylor and Francis. All have gone to great lengths to make this useful information available to a wider public.

Berne, 1 June 1993

Hans Hurni,
President, World Association of Soil and Water Conservation,
and Director, Group for Environment and Development,
Institute of Geography, University of Berne

Preface

This book was conceived on 30 June 1991, in the air above Siberia on our return from an expedition to the Altai mountains. Like many parents, we were aware neither of the implications of our decisions, nor of the eventual form of the resulting creation. As a mountain scientist with an interest in geographic information systems (GIS) and a GIS specialist with a strong interest in mountains, we decided to investigate the current state of the use of GIS in mountain areas around the world; and that the best way to do this was to have a meeting to bring people together and benefit from others' experiences.

In the process of locating many people with similar interests, we discovered that the World Association of Soil and Water Conservation (WASWC) was planning its second international conference on 'Environmental Regeneration in Headwaters', in Sec, Czechoslovakia, in November 1992. This seemed to present a good opportunity to hold our meeting, and also to introduce GIS to many researchers and land managers concerned with mountain regions from around the world. We are grateful to Martin Haigh (WASWC Vice-president for Europe) and Josef Krecek (Conference organizer) for agreeing to incorporate a one-day session on GIS into their conference, and for their organizational assistance. Ten presentations of GIS in mountain areas were given at the conference, and led to lively discussion of the potential and limitations of GIS. One of the official conclusions of the conference was that they 'are a technology with great potential for communication. Great care must be taken in activities such as interpolation, extrapolation, and classification. The power of the technology must not mask difficulties with data quality and understanding.' Many of these themes are echoed in this book.

This book is much more than the proceedings of the meeting in Sec, although it includes most of the presentations given there. Our purpose in assembling the book was to provide a broad overview of the applications of GIS in mountain regions through both reviews and case studies. We hoped to include papers on a wide range of themes, bringing together experiences from research and resource management institutions and the private sector from a variety of countries. We believe that we have succeeded in this; the papers describe case studies from five continents, considering topics from the traditional concerns of mountain scientists and resource managers to the increasingly critical issue of global climate change. We would not be so presumptuous as to say that the book describes the 'state-of-the-art', but we hope that it gives a flavour of the range of possibilities and challenges of using GIS in mountain areas. We anticipate that it will prove useful and interesting to many people who live, do research in, and manage mountain environments, whether they have experience in GIS or not. For those without such experience, we would recommend one of the many excellent introductory books (Aronoff, 1989; Burrough, 1986; Maguire *et al.*, 1991). Equally, we expect that many GIS specialists will find the book valuable, as many of the issues addressed in the papers are relevant for research and management in both mountain and other environments.

The book evolved gradually in mountain settings – including the English Lake District, the Slovak Tatra, and the Siberian Altai – and academic environments, particularly the Department of Geography at the University of Salford, where we worked together in 1991–92. As well as the authors of the papers, we must thank the many people who contributed to the birth and development of the book in various ways, including Yuri Badenkov, Sarah Cornelius, Jack Ives, Martin Parry, Jim Petch, and Victor Reviakin. We hope that they find their support for our integrating work has been fruitful. The introductory chapter benefited from the comments of David Favis-Mortlock, John Haslett, and Hans Schreier, as well as the GIS/cartographic expertise of Paul Brignall, who produced the map. We would also like to thank Richard Steele, of Taylor & Francis, for believing in the book and for being flexible with deadlines.

As far as we know, this is the first book to explicitly consider the use of GIS in mountains. As the papers show, this field is evolving rapidly, in many directions, and at a wide range of both spatial and organizational scales. We are sure that this evolution will continue, and hope that this book may assist in identifying some of the great potentials and challenges that exist and act as a catalyst for future work.

Martin F. Price
D. Ian Heywood

References

Aronoff, S., 1989, *Geographic Information Systems: a management perspective*, Ottawa: WDL Publications.

Burrough, P.A., 1986, *Principles of Geographical Information Systems for land resources assessment*, Oxford: Clarendon.

Maguire, D.J., Goodchild, M.F., and Rhind, D.W., 1991, *Geographical Information Systems: principles and applications*, London: Longmans.

Contributors

Richard J. Aspinall
The Macaulay Land Use Research Institute, Craigiebuckler, Aberdeen AB9 2QJ, UK

Ling Bian
Department of Geography, University of Kansas, Lawrence KS 66045-2121, USA

Daniel G. Brown
Department of Geography, Michigan State University, East Lansing MI 48824-1115, USA

Sandra J. Brown
Department of Resource Management Science, The University of British Columbia, #436E – 2206 East Mall, Vancouver BC V6T 1Z3, Canada

Bogdan Brzeziecki
Swiss Federal Institute for Forest, Snow and Landscape Research, Zürcherstrasse 111, 8903 Birmensdorf, Switzerland

David R. Butler
Department of Geography, University of North Carolina, Chapel Hill NC 27599-3220, USA

Kurt Culbertson
Design Workshop Inc., 120 E. Main Street, Aspen CO 81611, USA

Simon Ferrier
Armidale District, New South Wales National Parks and Wildlife Service, P.O. Box 402, Armidale NSW 2350, Australia

Sten Folving
Institute for Remote Sensing Applications, Joint Research Centre, 21020 Ispra, Italy

Susan Hall
Mount Revelstoke and Glacier National Parks, Canadian Parks Service, Revelstoke BC V0E 2S0, Canada

Patrick N. Halpin
Department of Environmental Sciences, Clark Hall, University of Virginia, Charlottesville VA 22903, USA

Bonny Hershberger
Design Workshop Inc., 120 E. Main Street, Aspen CO 81611, USA

D. Ian Heywood
Department of Environmental and Geographical Sciences, The Manchester Metropolitan University, John Dalton Building, Chester Street, Manchester M1 5GD, UK

Paula L. Horne
The Macaulay Land Use Research Institute, Craigiebuckler, Aberdeen AB9 2QJ, UK

Hans Hurni
Group for Environment and Development, Institute of Geography, University of Bern, Hallerstrasse 12, 3012 Bern, Switzerland

Suzanne Jackson
Design Workshop Inc., 120 E. Main Street, Aspen CO 81611, USA

Gavin Jordan
Department of Timber and Construction, The Buckinghamshire College, Brunel University, Queen Alexandra Road, High Wycombe HP11 2JZ, UK

Felix Kienast
Swiss Federal Institute for Forest, Snow and Landscape Research, Zürcherstrasse 111, 8903 Birmensdorf, Switzerland

Alexander Koshkariov
International Mountain Research and Information Laboratory, Institute of Geography, Staromonetny 29, Moscow 109017, Russia

Tatiana M. Krasovskaia
Department of Natural Resources Management, Faculty of Geography, Moscow State University, Moscow 119899, Russia

David R. Miller
The Macaulay Land Use Research Institute, Craigiebuckler, Aberdeen AB9 2QJ, UK

Jane G. Morrice
The Macaulay Land Use Research Institute, Craigiebuckler, Aberdeen AB9 2QJ, UK

Steve Mullen
Design Workshop Inc., 120 E. Main Street, Aspen CO 81611, USA

Heidi Olson
Design Workshop Inc., 120 E. Main Street, Aspen CO 81611, USA

Maria Luisa Paracchini
Instituto di Ingegneria Agraria, Università degli Studi, Via G. Celoria 2, 20133 Milano, Italy

José M.C. Pereira
Departamento de Engenharia Florestal, Instituto Superior de Agronomia, Universidade Técnica de Lisboa, Tapada de Ajuda, 1399 Lisboa Codex, Portugal

James R. Petch
Department of Environmental and Geographical Sciences, The Manchester Metropolitan University, John Dalton Building, Chester Street, Manchester M1 5GD, UK

Martin F. Price
Environmental Change Unit, University of Oxford, la Mansfield Road, Oxford OX1 3TB, UK

Ian Pulsford
Kosciusko National Park, New South Wales National Parks and Wildlife Service, Private Mail Bag, Cooma NSW 2630, Australia

Jörg Schaller
ESRI Germany, Ringstrasse 7, 8051 Kranzberg, Germany

Hans E. Schreier
Department of Resource Management Science, The University of British Columbia, #436E – 2206 East Mall, Vancouver BC V6T 1Z3, Canada

Andrew M. Stocks
GIS Unit, Department of Geography, The University of Salford, Salford M5 4WT, UK

Vladimir S. Tikunov
Department of Cartography and Geoinformatics, Faculty of Geography, Moscow State University, Moscow 119899, Russia

Cees J. van Westen
Department of Earth Resources Surveys, International Institute for Aerospace Survey and Earth Science (ITC), 350 Boulevard 1945, P.O. Box 6, 7500 AA Enschede, The Netherlands

Maria J. Perestrello de Vasconcelos
Departamento de Engenharia Florestal, Instituto Superior de Agronomia, Universidade Técnica de Lisboa, Tapada de Ajuda, 1399 Lisboa Codex, Portugal

Stephen J. Walsh
Department of Geography, University of North Carolina, Chapel Hill NC 27599-3220, USA

Otto Wildi
Swiss Federal Institute for Forest, Snow and Landscape Research, Zürcherstrasse 111, 8903 Birmensdorf, Switzerland

Guy Woods
Wildlife Branch, British Columbia Ministry of Environment, 617 Vernon Street, Nelson BC V1L 4E9, Canada

Bernard P. Zeigler
Department of Electrical and Computer Engineering, University of Arizona, Tucson AZ 85721, USA

1

Mountain regions and geographic information systems: an overview

D. Ian Heywood, Martin F. Price and James R. Petch

Introduction

Mountains are distributed across all of the world's continents. They include a vast diversity of environments: from the wettest to the driest; from hot to cold; and from sea-level to 8848 m at the top of Mount Everest (also known as Sagarmatha and Chomolungma). The cultural diversity found in mountain regions is also great. This is a reflection both of the variety of environments, which permit diverse means of livelihood; and of the importance of mountains as places of refuge, security and recreation – both spiritual or physical – often at the margins of politically and topographically-defined territories. The diversity, marginality, and strategic importance of mountains, together with vastly different rates of change in different components of their physical, biological, and societal systems, present great challenges for the use of geographic information systems (GIS).

The purpose of this chapter is to provide an overview of the current applications, issues, and challenges that are receiving attention by mountain scientists using GIS. Beginning with a consideration of the special characteristics of mountain regions, it examines the extent to which the use of GIS in these regions reflects such characteristics, introduces the other chapters in the book, and assesses whether current GIS technology can meet the demands of scientific research and management in mountain areas.

In contrast to the widespread use of GIS in other areas, the technology has received relatively limited use in mountain environments. This must, however, be placed in the broader context of a general naivety towards the science and management of mountain ecosystems. Ives (1992: xiii), for example, suggests that to the majority of people, mountains have 'the appearance of being remote, durable and barely affected by the environmental ills that beset the more densely populated areas of our global living space'. Therefore, funding for the development of appropriate strategies to improve the use and management of these regions is often a low priority. However, in recent years, growing concern over the environmental degradation of mountain ecosystems has meant that mountain issues are gradually finding their way onto environmental and political agendas. An example of this growing interest

was the formulation of a Mountain Agenda for the United Nations Conference on Environment and Development (UNCED), held in Rio de Janeiro
during June 1992, with the goals of:

- making an authoritative statement on the environmental status and development potential of the world's mountains;
- disseminating this information in the widest possible form;
- publicizing the urgent need for priority ranking of the mountain problem;
  and
- providing some guidelines for a practical response to the problems and
  challenges of the mountains for consideration by world leaders (Ives, 1992).

All of these broad goals require information, as recognized in Chapter
13, on 'managing fragile ecosystems: sustainable mountain development,' of
Agenda 21, the programme for action resulting from UNCED (Quarrie, 1992).
Yet, at the present time, the necessary information either does not exist or
is far from comprehensive in its coverage. In this respect, the Mountain Agenda
and the consequent Chapter 13 of Agenda 21 can be seen as an excellent
basis for arguing for the wider use of GIS to assist in developing our understanding and improving our approaches to the management of mountain
environments.

Several organizations at regional, national, and global scale have already
recognized this potential and are developing long-term monitoring programmes
in which GIS will play a central role. The chapters in this book provide a
range of examples, from the globe (Halpin, 1994), to whole nations
(Koshkariov *et al.*, 1994), states within nations (Pulsford and Ferrier, 1994),
administratively-defined areas, such as national parks (Schaller, 1994; Walsh
et al., 1994), and smaller study areas, defined for specific research projects
(Paracchini and Folving, 1994) (Figure 1.1). In all cases, the objectives are to
record baseline information about the state of mountain regions and the
anthropogenic stresses that affect them; to evaluate methods for the continued monitoring and sustainable development of these environments; and, in
some cases, to evaluate future scenarios deriving from the interaction of
biophysical and societal processes, usually through simulation models
(Brzeziecki *et al.*, 1994; Halpin, 1994; Schaller, 1994; Vasconcelos *et al.*, 1994).

It needs stating from the outset that there is nothing unique about the
character of GIS applications in mountain areas. This is to be expected, since
although we can regard mountains as distinct areas of the earth's surface with
a common set of issues, they are not unique. In fact, these issues, including
resource assessment; hazard prediction; environmental impact assessment;
and ecosystem management, are determined not by the nature of any particular landscape but by society's collective objectives. Nevertheless, the use
of GIS in mountains requires some special considerations. These derive from
the particular characteristics of mountain environments and the peculiarities
of data collected in these complex, dynamic regions, and also from the

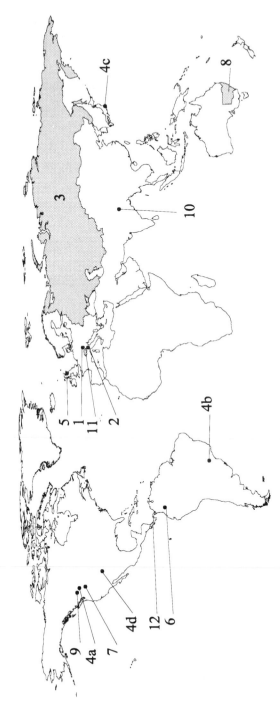

Figure 1.1 Locations of GIS study areas described in the chapters of this book.
1. Berchtesgaden National Park, Germany (Schaller); 2. Val Malenco, Italy (Paracchini and Folving); 3. Commonwealth of Independent States (Koshkariov, Krasovskaia, and Tikunov); 4. a: Bow-Canmore area, Canada; b: Tamandua region, Brazil; c: Nishi-Tama region, Japan; d: Colorado, USA (Culberson et al.); 5. Cairngorms, Scotland (Miller et al.); 6. Rio Chinchina catchment, Colombia (van Westen); 7. Glacier National Park, Montana, USA (Walsh et al.); 8. New South Wales, Australia (Pulsford and Ferrier); 9. Mount Revelstoke and Glacier National Parks, British Columbia, Canada; 10. Sagarmatha National Park, Nepal (Jordan); 11. Switzerland (Brzeziecki, Kienast, and Wildi); 12. Costa Rica (Halpin).

limitations, originating in current GIS software and ecological theory, on our ideas about how mountain systems work.

The great variability of mountain environments at all spatial and temporal scales requires great care, both in the choice of scales for data collection and storage, and in extrapolations from locations with high-quality, long-term data to data-poor areas, even over quite small distances. Methods of assessment and extrapolation developed in lowland environments may be quite unsuitable for heterogeneous, fast-changing, mountain environments. When applied in these areas, the assumptions embedded in such methods must be made explicit and assessed for their applicability for a specific task, using detailed local knowledge. Further problems include:

- modelling extensive variations in relief and the shape of terrain with a technology that essentially remains two-dimensional:
- ground-truthing remotely-sensed, or model-generated, data in isolated environments;
- handling the temporal aspects of rapid environmental change; and
- communicating the results of GIS-based analyses to local populations hampered by lack of understanding of such new technologies.

The special characteristics and complexity of mountain areas

Mountains and uplands comprise about one-fifth of the world's terrestrial surface, and are directly or indirectly important for more than half of the world's population (Ives, 1992). They are home to about one-tenth of the world's population; supply natural resources – food, wood, and minerals – to an even greater proportion of the global population; are at the upper end of most of the world's river catchments, providing water, nutrients, and energy to those living both nearby and at distant locations downstream; provide environments for recreation and tourism for visitors from both nearby and far away; include centres of biodiversity and refugia for relict species and communities; and are of great spiritual and aesthetic significance to many people.

Although there are numerous definitions of what mountains are, there is general agreement that they are areas of high relief. They range from isolated peaks and islands to complex systems of considerable extent (Gerrard, 1990). Thus, the physical characteristic that best defines mountains is their three-dimensionality, which produces contrasting environments at different elevations. Superimposed on this altitudinal zonation, however, are variations that derive from the aspect, slope, and topography of a particular mountain or region (Barry, 1992b). In fact, it is this three-dimensionality that poses the greatest challenge for modelling these regions using GIS, for the simple reason that most GIS and the data they incorporate still treat the world as if it were flat. Nevertheless, GIS used in mountain areas usually incorporate digital terrain models, which permit the representation of the three-dimensional nature of mountains (Stocks and Heywood, 1994).

The importance of the third dimension can be summarized by a consideration of global and local climate. In the northern hemisphere, south-facing mountain slopes receive more radiation than north-facing slopes. Consequently, the former tend to have a longer snow-free period, larger ranges of temperatures at diurnal and annual scales, and drier micro-climates. The degree of these differences also varies with slope angle and latitude. The topography of mountain regions has further effects on local and micro-climates both in valleys and on upper slopes and summits. In valleys, cold air collects during stable meterological conditions, usually experienced in winter. In such 'inversion' conditions, the typical decrease of temperature with increasing altitude does not apply: upper slopes are far warmer than valley bottoms and lower slopes. However, upper slopes and summits tend to be especially harsh environments, characterized by strong winds, low temperatures, and moisture deficits. Consequently, a GIS is valuable because, despite its two-dimensional view of the world, it is still possible to model such features as aspect, slope, and height.

The ability of a GIS to model the nature of the terrain is essential because the local and micro-climatic variations described above strongly influence the biophysical components of mountain environments: air; water in its different phases; soils; vegetation; fauna. Precipitation and water storage and flow all vary with altitude and aspect. Thus, for example, in the northern hemisphere, mountains may have remnant glaciers and permanent snowbeds on their north-facing slopes, but none on their south-facing slopes. Both soils and vegetation tend to occur in belts at different altitudes, their concentricity around a mountain or range offset by the variations deriving from differences in local and micro-climate which result from aspect; for instance, the tree-line is typically lower on north slopes in the northern hemisphere.

In addition, variations in local and micro-climates also influence patterns of housing, agriculture, and recreation. Valley floors are often particularly desirable locations for settlement and agriculture. However, the stagnation of cold air, which can result in the concentration of airborne pollutants from sources as diverse as wood-fires and automobile exhausts, means that winter resorts are often developed on upper slopes. These may replace small high settlements, previously used only by herdsmen bringing animals to pastures during the brief summer season. This pattern of resource use, which permits domesticated animals to benefit from a range of food sources in different seasons, is found in many mountain regions.

To develop an understanding and appreciation of the optimal locations for settlements, it is also essential that the GIS should be able to identify areas affected by the likelihood of natural hazards, including avalanches; rockslides; floods; and forest fires, whose distribution is influenced by complex interactions between local climates, human activities, and soil, bedrock, and vegetation characteristics (Hewitt, 1992). While likely occurrences of some of these hazards are somewhat predictable, those of others – such as mass movements triggered by earthquakes and volcanic eruptions – are less so, although they may have far more extreme effects than more frequent events. This is not simply a mapping problem and a GIS must be capable of addressing the issue of how, when, and where these high-risk events are likely to occur.

In the national and global context, a further important characteristic of mountain environments is that they tend to be marginal areas which are physically and/or culturally distant from centres of political power, and directly contribute little to national economies. Mountain agriculture and forestry are often seen as a drain on national and even regional budgets although they provide both direct and indirect benefits for considerable numbers of people. In general, per capita levels of investment tend to be far lower in mountain areas than in adjacent lowlands, and investments (e.g. large hydroelectric schemes and tourism developments) may not directly benefit long-established mountain dwellers, who may even be disadvantaged or displaced by such projects. Placed in a GIS context, the marginality of mountain environments means that there is a need for a tool to assist policy makers, planners and environmentalists at local, regional, and national levels to develop strategies for the economic and ecological management of these regions.

The geographical and economic marginality of mountain areas is often exacerbated by their political sensitivity. Many mountains form natural barriers, so that national boundaries (often in dispute at the present or in the past) run through them. Thus, the sensitive nature of these border areas and, not infrequently, armed conflict, may greatly hinder the collection of essential data through air- or spaceborne remote sensing and fieldwork (Schweinfurth, 1992).

Developing and using GIS in mountain areas: data issues

Potential and actual users of GIS in mountain environments are confronted by a wide range of issues with regard to the data that can be incorporated and used. The availability and collection of data may be affected by historical, political, climatological, topographic, and many other considerations. Equally, special attention must be given to the processing, analysis, and modelling of data relating to mountain environments because of their spatial and temporal complexity. These issues are discussed below.

Regional diversity

Superimposed on the great physical and biological heterogeneity of mountain regions are diverse histories and patterns of human use. When developing a GIS, this is important, since the patterns and histories of human use of mountain regions greatly affect the types and availability of existing data, and may restrict the use which can be made of them. A useful typology in this respect is provided by Grötzbach (1988), who differentiates between young and old mountains. The former are the relatively sparsely-settled mountains colonised by the Europeans in recent centuries. Examples include the mountains of North America, Australia, and New Zealand. Here, perhaps because of the introduction of European concepts of property rights and a market-oriented approach to development, there is very often a wealth of geo-referenced information suitable for use in GIS. It usually exists in the form of paper maps, land inventory surveys, and property ownership records. More recently, the application of remote sensing has led to the development of more regular broad-based land inventories suitable for use in GIS.

Most of the old mountain regions are relatively densely settled. They can be divided into three categories. The first of these is characterized by a decline in traditional agriculture and forestry, linked to depopulation except in areas experiencing a growth in tourism, which in many cases has become the basis for the economy. These regions are typical of many of the mountains of western Europe. The long human history of these regions means that they too are often data-rich in terms of information suitable for inclusion in a GIS. Spatial information is often available at a much more detailed scale than for young mountain regions, and the temporal range is generally much greater.

The second category of old mountains includes most in less-developed nations. Traditional subsistence and/or herding agriculture is the dominant land use and there is a tendency towards over-population. Most of these regions are data-poor, with the exception of those which found themselves subject to colonialism and where detailed mapping was undertaken. Thus, the majority of available spatial information tends to have been derived from satellite surveys in recent years; it is very often owned by the military. The political instability of many of these regions makes access to data particularly problematic.

The third category of old mountains has, until recently, largely been characterized by collectivized or nationalized agriculture and forestry. These include the Carpathians and the mountains of the former Soviet Union and China. With the fall of the communist regimes in all of these regions except the latter, their social, economic and political structures are now in a state of flux, as are the availability of, and access to, geo-referenced information. While it is very hard to make any general statements, two examples may be given. First, preliminary work on GIS in the Altai mountains of Siberia (Carver *et al.*, 1993) has revealed a plethora of geographic information that is as extensive in its coverage as that for many of the young mountain regions.

Second, the Atlas of the Tatra National Park in Poland (Trafas, 1985), with a remarkable array of thematic maps, many of them at scales of 1:25 000 or better, is probably unmatched even in western Europe. Both of these comprehensive data sources are ripe for inclusion in GIS. Nevertheless, in general, scientists in this third group of old mountain regions not only suffer from a lack of hardware, software, and training, but frequently face the additional problem that existing paper maps often contain intentional distortion and misleading generalization of key geographic features such as roads, towns, and rivers, most probably carried out for security reasons.

Data sources

Three main sources of data have been used in the development of GIS for mountain areas: maps; field survey data; and remotely-sensed imagery from aerial photography or satellites. Such imagery is typically used to provide a basic classification and inventory of land use/cover types, as described, for instance, by Paracchini and Folving (1994) and Walsh *et al.* (1994). However, the availability of such classifications varies greatly between mountain regions. Aerial photography may not be available or possible, particularly in areas which are politically sensitive; and digital remote sensing data may be of insufficient spatial and/or temporal resolution. In addition, mountain areas have strong topographical gradients which markedly affect reflectance patterns; and the classification of large areas is often hindered by frequent and widespread cloud cover.

For any combination of these reasons, full coverage of an area may require the collection of images over many days or even seasons. This may only be a partial solution since, because of the rapid phenology of many mountain plant communities and crops, images of an area obtained only a few days apart may have significantly different reflectance characteristics. Time of day is also a critical constraint: even at solar noon, shadows on steep slopes can make classification very difficult. Finally, short snow-free seasons limit the period during which useful imagery can be obtained. While air- or spaceborne remote sensing can be of great value in developing land cover/use classifications of mountain areas, fieldwork remains essential and this, again, may be hindered by political considerations.

Enhancement and classification of satellite images require the control of topographic effects. This is best achieved using a GIS to isolate these effects of aspect and gradient (Paracchini and Folving, 1994). Global Positioning Systems (GPS) have recently been used with GIS to detect sites in mountain areas which are used in image classification (Rodcay, 1991), and can be used to determine sampling positions and heights to create grids for photo rectification (Haefner and Hugentobler, 1985). Such enhanced images have been used for many applications of GIS in mountain regions, including resource evaluation (Walsh *et al.*, 1990), land cover mapping (Downey *et al.*,

1990), biodiversity analysis (Rodcay, 1991), and assessment of forest fire effects (Butler *et al.*, 1990).

Data models for mountain environments

One necessity in the development of a GIS for a mountain area is the ability to model its three-dimensional complexity. From the literature, two broad approaches can be recognized. The first involves the use of digital terrain models (DTMs), also known as digital elevation models (DEMs), to provide a categorization of zones or elements of a mountain area according to slope, elevation, and aspect. The second uses landscape units, constructed from a synthesis of environmental data, which reflect the character (structure, sustainability, and responsiveness) of an area rather than its physical form (height, shape, and exposure) alone.

The DTM approach

The DTM provides a framework for the use of GIS in mountain areas by capturing the three-dimensionality of mountain environments as a two-dimensional mosaic. This can be constructed of either regular or irregular tessellations, using the grid or triangular irregular network (TIN) methods, respectively. For each tessellation, the three major terrain elements of height, aspect, and slope angle can be recorded as attributes (Weibel and Heller, 1991). The spatial relationships between these attributes are then used to portray the three-dimensional character of the area. Given the critical importance of terrain in understanding mountain environments, DTMs are an essential element of GIS for mountain areas. A separate overview chapter (Stocks and Heywood, 1994) looks at the principles and applications of DTMs in mountain environments.

The landscape approach

The landscape approach does not deal directly with the problem of capturing the physical shape of terrain. Instead, it incorporates higher levels of knowledge about the character and behaviour of mountain systems. The approach draws on the principles of landscape ecology (e.g. Forman and Gordon, 1986). As a first step, a landscape model of the unchanging physical characteristics of an area is contructed from ground survey information, aerial photography, and base maps. These include relief maps, from which facets of the landscape are defined, each with a specific combination of altitude, aspect, and slope. In practice, these maps are drawn by hand, and include a limited number of categories decided by the cartographer. Other data on geology, soil types, hydro-meteorological conditions, and vegetation are used to produce a typology of the landscape units that represent permanent conditions of the landscape (Bucek and Lacina, 1981).

While the development of such landscape maps has, to date, been time-consuming because of the need for expert knowledge, new methods using knowledge-based systems are being pioneered (Burkmar *et al.*, 1992). These landscape maps are then used in conjunction with land cover or land use maps, or satellite remote-sensing products (Downey *et al.*, 1991), to assess the appropriateness or stability of use in each land parcel in relation to the natural conditions. Such assessments have two parts, considering both the suitability of a site and its position in the two-dimensional structure of stable and un-stable sites (Petch and Kolejka, 1993). One basic problem of the landscape approach is that assessments of stability, generated by the overlay of landscape units, are not easily interpreted by non-experts. This presents particular problems in the use of this information in policy-making. However, the landscape approach does represent an alternative method of incorporating knowledge of the complex nature of mountain areas into GIS.

Data quality and GIS techniques

Many of the problems hindering the application of GIS in mountain environ-ments relate to the quality of the available data and their suitability for use in GIS. Data for almost all aspects of mountain ecosystems are very hetero-geneous in their length and frequency of record, spatial coverage, and avail-ability. Three examples of data issues, each with an increasing degree of challenge posed by the problems deriving from heterogeneity, are discussed below: overlay techniques; process modelling; extrapolation.

Overlay techniques are perhaps the most frequently used to obtain additional information from existing data sets. The problems engendered by overlaying maps of different scales and/or based on various coordinate sys-tems are well-known in all environments. However, the three-dimensional complexity of mountains can exacerbate these problems to an extreme extent. One example is provided by Beissmann (1989) who attempted to create a GIS of the Wallackhaus area of the Hohe Tauern National Park, Austria, using a number of high-quality thematic maps, all based on detailed field research in the 1970s and earlier. However, they did not have a consistent, accurate cartographic base and were at different scales. As a result, there was considerable disagreement in the location of land-cover categories; even those that should have been well-defined and consistent between maps, such as talus slopes. Such problems may become less critical as more spatially-consistent technologies become available. However, great care must be taken to use these effectively; and both fundamental issues and time-consuming problems will still remain when using data sources that are uncorrected or have different scales, resolutions, and degrees of generalization.

With respect to the modelling of environmental processes, it is essential to correctly choose both the appropriate variables and the temporal and spatial scales at which they are defined. An example is provided by Wadge

(1988), who has shown how the accuracy and quality of a single data set, in his case a DTM, can influence the results of slope process models. The research showed that most DTMs are inappropriate for determining the dynamics of slope stability where the value of every grid cell is crucial to the model. Similarly, dynamic slope-flow models fuelled with DTM data of the wrong spatial resolution may lead to a magnification of the scale of the problem, so that material is sent down the incorrect path. Nevertheless, models of slope energy balance appear to be less sensitive to individual pixel values, so that it is quite plausible that DTMs derived from topographic maps at scales greater than 1:1000 may provide an overview suitable for the development of zone mapping. In general, the choice of (spatial and temporal) scale of the data is critical for successful, reliable process modelling; data whose resolution is either too great or too small may provide results that are incorrect, misleading, or even meaningless and often untestable.

Extrapolation is an activity for which GIS are widely promoted and often used. As noted by Halpin (1994) in this volume, this can be very risky without a deep understanding of the inter-relationships of the variables involved in the extrapolation, and the risks of inaccurate or improbable extrapolation increase with the number of variables and the complexity of the environment. As Haslett (1994) has recently shown using fractal analysis of the Berchtesgaden GIS (Ashdown and Schaller, 1990), mountain environments are the most spatially complex. Thus, for any phenomenon, extrapolation can be highly questionable, even between locations with reliable data.

Mountain climates provide many examples of complex, dynamic phenomena (Barry, 1992a). They are characterized by marked diurnal and seasonal cycles, with high variability at all spatial scales. As mentioned above, climate is a critical controlling factor in the distribution of environmental components and the uses of mountain environments, so that the possibility to extrapolate, in both time and space, from existing information is highly desirable and can theoretically be achieved using GIS. Yet, given the great variability of climates at all temporal and spatial scales, the useful characterization of the climate of any mountain area requires long-term records from dense networks of stations at a wide range of altitudes and on slopes of different aspects.

Unfortunately, such records are rare. Climatic data tend to be more readily available from settlements close to the mountains or in valleys, rather than from mountain sides and summits. The length of records is often inadequate to provide a realistic assessment of the mean and, especially, the extreme values of relevance to both human beings and other components of mountain ecosystems. This problem is exacerbated by changes in the location of recording stations, a critical problem in areas with large variation over small distances; or because data collection is only seasonal; for instance, in winter to provide information for predicting avalanches or spring runoff.

The conclusion with respect to this and the preceding data issues is that the use of GIS must involve awareness of the limitations of not only the available data but also the understanding of environmental processes and the technology in use. Given such awareness, GIS can be a valuable technology for descriptive, analytical, and evaluative purposes.

GIS Applications: from inventory to management

GIS have been used for many objectives in mountain areas around the world. To all intents and purposes, these applications reflect environmental, cultural, and economic issues that have come to the fore in recent years including, for example, forestry; mining; ecology; tourism; hazard mapping; visual impact assessment and climate change modelling. These broad application areas have seen the use of GIS in two main ways: first, as tools to assist in resource inventory and the integration of data and, second, as a mechanism for analysis, modelling, and forecasting to support decision-making (McKendry and Eastman, 1990). In many cases, the GIS is used in both capacities. The following sections present four categories of GIS applications: regional resource inventory and planning; evaluation of natural hazards; research and resource management in and around protected areas; and simulation and prediction, into which the papers in this book have been divided, and identify key points in these papers. Many other studies are also cited for illustrative purposes. However, the following sections should not be construed as an exhaustive review of a rapidly-evolving field whose literature is widely scattered and often rather inaccessible.

Regional resource inventory and planning

Data inventory and integration has been the most consistent use of GIS since the development of the first GIS for mountain areas – such as the Lluçanes area of the Spanish Pyrenees (Alegre, 1982) and the Grindelwald area of the Swiss Alps (Steiner and Zamani, 1984) – on mainframe computers. Most mountain GIS have been developed with data at a limited range of scales for specific purposes in well-defined areas. While a few attempts have been made to create multi-purpose data banks or databases for mountain areas, there have been considerable difficulties in designing such versatile GIS. One example was the creation of a GIS for Nepal, through a joint project involving the Global Resource Information Database (GRID) of the United Nations Environment Programme (UNEP), in collaboration with the International Centre for Integrated Mountain Development (ICIMOD), in Kathmandu (Simonett, 1993). The objectives of this project include the creation of environmental databases at different scales and resolutions: national (1:1000 000–1:3000 000), zone district (1:50 000–1:500 000), and local (1:10 000–1:50 000). In fact, the local and zonal GIS are partial in coverage and oriented towards particular objectives. There is, nevertheless, a general attempt to approach the problem of a central GIS resource.

One example of a smaller scale of operation is the design of a multi-user GIS for the Zdar Hills protected area of the Czech Republic (Pauknerova *et al.*, 1992). This is a GIS with over 20 data layers, intended for use by different groups and agencies (Downey *et al.*, 1990). The design of the system includes two specific elements which permit its use for particular purposes: a system of land units, and a territorial system of ecological stability. Using ideas of landscape ecological management (Michal, 1992), these provide the methodological basis for using the same data layers for different purposes. The GIS can therefore be used as a general tool for data integration. Similar work has been done in the Malcantone region (65 km^2) of Switzerland, in an intensive study of landscape change and landscape protection (Haefner and Hugentobler, 1988; Haefner *et al.*, 1991). In this project, the analysis of landscape, using remote sensing and GIS, is the basis for studying economic and environmental problems, such as conflicts between land uses with respect to landscape protection, and the identification of areas at risk from flooding, in order to develop appropriate land use planning and management strategies.

In this volume, Schaller (1994) describes the conceptual bases for a very detailed GIS for the management of Berchtesgaden National Park and the neighbouring region in Bavaria, Germany. The chapter is unusual in that it presents the ecological theories on which the development of the GIS was based; most GIS are initially created as tools for data integration and analysis without a theoretical framework. The approach developed in Berchtesgaden has since been applied to other mountain areas in China, Kenya, and New Zealand (Ashdown and Schaller, 1990). Paracchini and Folving (1994) present the initial development of a GIS for regional planning and modelling in the Italian Alps. The principal objective of the GIS will be to provide low-cost, accurate mapping and methods for monitoring environmental conditions, with particular regard to hydrological resources.

Also in this volume, a number of GIS studies in their early phases are described by Koshkariov *et al.* (1994) at scales from the entire Commonwealth of Independent States to individual *rayons* (districts). At present, the main aim of this work is the generation of digital atlases which, in the future, will be used for regional planning. Pulsford and Ferrier (1994) describe the implementation of a GIS to support the multiple responsibilities of the National Parks and Wildlife Service (NPWS) of New South Wales, Australia, both in National Parks and throughout the state. The approach is notable for its decentralized approach and the use of low-cost equipment. The GIS has been used to assist in a wide range of management and research activities.

Within more narrowly defined objectives, GIS have been created as data integration tools for many mountain areas. Among the objectives of a project on forest resources in Nepal were the documentation of historic change, the determination of relations between use and site factors, and the prediction of future resource use (Schmidt and Schreier, 1991). GIS appears to be playing a key role in wider strategies of resource and environmental management in Nepal (Schreier *et al.*, 1990, 1993). A similar GIS strategy has been developed for the Great Smoky Mountains National Park, USA (Parker *et al.*, 1990), where the GIS is used to analyze the spread of forest diseases and to produce inventories of forest types and biodiversity.

One of the significant potentials of GIS is to use their three-dimensional and graphic capabilities to make inventories of not only the usual biophysical and cultural resources, but also visual resources. Thus, current landscapes can be portrayed and the effects of future planning scenarios assessed. In this volume, Culbertson *et al.* (1994) present a number of case studies in which visual resources were considered as an integral component of the regional resource base when developing plans for the future of mountain areas – with recreational facilities, settlements, transportation corridors and, in two cases, mines – in Brazil, Canada, and Japan. The methodology proved valuable both as a means for integrating a wide range of information and also as a basis for public involvement in the planning process. Their final example, from Colorado, shows the potential of GIS as an integrative tool for planning the use of multi-jurisdictional resources. Miller *et al.* (1994) also focus on the assessment of recreational resources through the evaluation of scenery and the visual impacts of land management practices, particularly forestry, in the Cairngorm mountains of Scotland.

As shown by some of the examples above, the greatest and most interesting challenge for mountain scientists is the use of GIS technology in an investigative role: seeking answers to 'what-if' questions such as: which areas are at risk from landslides?; what will be the economic and cultural implications of particular tourism/agricultural activities?; how will deforestation effect soil erosion?; or, what are the potential implications of long-term environmental change? To answer these questions, GIS can provide a wide variety of analytical, modelling, and forecasting modes. While resource inventory is the essential first step, the greatest potential of GIS technology is its use to link resource data with explanatory and predictive models and expert knowledge. The following sections consider three general themes: natural hazard evaluation; research and resource management for conservation; simulation and prediction, recognizing that any GIS may have the potential to be used for any or all of these purposes and for many others.

Evaluation of natural hazards

Given the dynamic nature of mountain environments, natural hazards, such as avalanches, landslides, fires, and floods, are major topics of concern for management. GIS can be used in many ways to increase our understanding of these phenomena and assist in developing adaptation and prevention strategies. Two chapters in this volume concentrate on natural hazards. Van Westen (1994) provides an overview of the use of GIS for landslide hazard zonation, using examples from the Andes of Colombia. A range of analytical approaches are presented and evaluated with reference to their applicability with respect to different levels of data availability and scales of analysis. In contrast to this classical approach based on the integration of remotely-sensed imagery, maps, fieldwork, and expert knowledge, Vasconcelos *et al.* (1994) present an approach to fire spread modelling in which a GIS is linked to a knowledge-based simulation model. This work is at an early stage of development, but shows considerable promise for developing predictive understanding of discontinuous processes of which fires are but one example in mountain environments.

The use of GIS to help in modelling the dynamic nature of mountain ecosystems is also characterized by the work of Walsh *et al.* (1994) who have been involved in producing a probability map of where, and on what terrain, avalanches are likely to occur in Glacier National Park, USA. Their multi-criteria approach weights the importance of different data layers (height, aspect, and slope angle) within the GIS relative to their importance in the avalanche formation process. The study is a clear example of how a GIS may provide valuable information for the modelling of physical processes. Both Walsh *et al.* (1994) and Pulsford and Ferrier (1994) also consider aspects of fire management, from the analysis of past events to potential assessment and suppression. Comparable work has been undertaken by van Wagtendonk *et al.* (1990) in Yosemite National Park, USA.

Research and resource management in and around protected areas

GIS has been widely used as a tool to aid the design, management, and monitoring of National Parks and other protected areas (Willison *et al.*, 1992). Numerous case studies reveal the benefit of applying this technology for both policy and research.

One of the major programmes of research in a National Park began in the early 1980s in Berchtesgaden National Park, Germany (Schaller, 1994). Part of this programme was the development of a zoological GIS which includes topographical data, habitat information derived from literature reviews and fieldwork, and the results of field surveys (d'Oleire-Oltmanns and Franz, 1991). The system has been used for habitat analysis and space distribution models of both individual species (d'Oleire-Oltmanns, 1987; Schuster, 1990) and species assemblages of vertebrates and invertebrates (d'Oleire-Oltmanns, 1990; Haslett, 1990) and for simulating changes in populations (d'Oleire-Oltmanns *et al.*, 1991). The visual outputs which derive from this work are used for both research and resource management. A GIS has also been used for zoological research in Langtang National Park, China, where the locations of red pandas, recorded using GPS technology, have been compared with habitat evaluations for different seasons. The resulting maps have been used to estimate populations and develop management plans (Yozon *et al.*, 1991).

Another theme of research has been the monitoring of landscape change.

One major study considered the twelve National Parks of England and Wales, most of which are in upland or mountain areas. The study, based on multi-temporal aerial photography and field surveys, showed how GIS could assist in developing a methodology for time-series analysis of complex data sets (Bird and Taylor, 1990). Satellite remote sensing images have also been used as a multi-temporal data source, combined with existing base maps and maps deriving from field research, for determining management regimes in Mercantour National Park, France (Claudin *et al.*, 1993).

This volume includes five chapters which specifically consider National Parks and other protected areas, as well as adjacent land. Walsh *et al.* (1994) present a summary of a major ongoing research programme conducted over many years in Glacier National Park, USA. Building on considerable previous research in the Park, the programme has included four stages. Beginning with the creation of an integrated database from a wide variety of sources, the work proceeded to the evaluation of various landscape components, such as avalanches, wetlands, and areas affected by fire. This led to the analysis of scale dependencies and spatial phenomena, which is vital both for understanding ecosystem processes and for structuring the GIS. Current work focuses particularly on the construction of models of alpine treeline, especially in relation to topoclimatic factors and snow cover. Pulsford and Ferrier (1994) present a complementary chapter which focuses on the implementation of a decentralized approach for the use of GIS by the National Parks and Wildlife Service of New South Wales, Australia. Case studies include the use of GIS in fire management and planning, surveys of fauna, wilderness assessment, archaeological studies, and the development of wildlife habitat links through areas proposed for logging.

The last of these topics indicates one of the major potentials of GIS: its usefulness as an objective methodology for developing management strategies in both protected areas and adjacent lands. The integrative value of GIS is especially valuable for resources and processes that cross the human-defined boundaries of protected areas: for instance, fires, animals, and people. Brown *et al.* (1994) give a concise account of the use of GIS for providing the information necessary to help resolve conflicts between forestry and wildlife conservation interests in an area that includes logging leases and two National Parks in the Rocky Mountains of British Columbia, Canada. They show that overlay techniques in a GIS may be used to define population goals and habitat requirements for a viable caribou herd and develop trade-off scenarios for the modification of wood harvesting schedules. Thus, the GIS may be used as a neutral system for identifying potential conflicts through the development of scenarios for each of the stakeholders. This work has recently been extended to the economic evaluation of trade-offs between conserving forests for logging or as caribou habitat (Thompson *et al.*, 1993). Such multiple accounts analysis is a major potential use of the data in GIS for economic decision-making which is just beginning to be realized.

Jordan (1994) also considers potential conflicts between different groups:

in this case, the management personnel and villagers living in Sagarmatha National Park, Nepal. His study of deforestation risk suggests that some of the aspects of the theory of Himalayan degradation (Ives and Messerli, 1989) have been overstated. His simple model of deforestation risk, initially developed in England, was tested in the field, leading to an improvement in predictive value through fieldwork and the application of local expert knowledge. This shows the importance of field testing of any results generated by deterministic GIS models, which are often presented, and even used for policy development, without field verification.

Simulation and prediction

The use of GIS to suggest the future of managed and unmanaged mountain environments requires both the inclusion of the necessary range of variables at suitable spatial scales and a detailed understanding of their interactions. However, for many of the questions for which scientists and managers require answers, neither the data nor the theory are adequate. Thus, GIS may be conceived as systems that are valuable not merely for recording, managing, and integrating data, but also for generating ideas and exploring relationships.

The use of GIS for simulation and prediction is, at present, not very well-developed, especially for highly complex environments such as mountains, although many possibilities exist. An early use of simulation modelling for a small area (100 km^2) was undertaken in the Swiss valley of Davos in the early 1980s (Binz and Wildi, 1986). In this work, a highly detailed GIS, with 50 m × 50 m grid cells, was used to evaluate the effects of various land-use scenarios, such as increased tourist development, or the maintenance of an attractive cultural landscape, on avalanche hazards as well as wildlife habitat, vegetation types, and other topics of interest for nature conservation.

One area of particularly great potential value is to use GIS to look back to the past and forward to the future. Such sequential analysis has been undertaken in the Middle Mountains of Nepal, showing a change from deforestation in the late 1940s to the 1960s, to successful afforestation in the 1980s (Schreier *et al.*, 1993). Over the same period, agriculture has been expanded onto more marginal sites. The combination of these two processes is leading to deficits in food sources for grazing animals. By incorporating a microclimatic model and expert knowledge into the GIS, the work has allowed the definition of the best uses for land in order to optimize the production of both trees and fodder. Such approaches might also be used for other diverse temporal processes, such as mass movements, the expansion of recreational impacts, and changes in the locations of glacier margins, permitting the simulation of ice dynamics.

A number of the chapters in this volume use various approaches to simulation, from simple map overlay models to decision trees and expert systems

(Vasconcelos *et al.*, 1994; Walsh *et al.*, 1994; Pulsford and Ferrier, 1994). However, the two papers which focus specifically on modelling and prediction of real mountain environments both consider what may be the most important long-term issue which the people and managers of mountain environments will have to consider: global climate change.

Brzeziecki *et al.* (1994) describe a major research project to simulate the impacts of climate change on the vegetation of Switzerland, a country that is nearly three-quarters mountainous. The project began with the integration of a DTM and environmental data sets as the basis for developing a model of potential natural vegetation on a 250 m × 250 m grid for the entire country. After quantitative and qualitative comparisons of the results of the model with 'reality' (as represented on vegetation maps) showed a reasonable degree of reliability, experiments were conducted to assess likely effects of increasing temperature. This led to plausible results, with forest communities 'moving' upwards as climatic conditions become more favourable.

Halpin (1994) presents a similar approach for the vegetation of Costa Rica, albeit at a coarser spatial scale, and also presents a comparative study of the impacts of climate change on the vegetation of a hypothetical mountain digitized into GIS of Alaska, California, and Costa Rica. He notes that previous approaches to assessing the effects of increasing temperature on ecological zones in mountains used very simplistic rules, rather than GIS approaches based on detailed understanding of ecoclimatic relationships for asymetric zones. The conclusions of past studies may therefore be not only simplistic but inaccurate; an important concern when they are applied to planning for protected areas, ecosystems, and species. The chapter concludes that realistic plans for the management of mountain ecosystems require the integration of GIS analysis and evolving ecological theory, with validation through fieldwork and remote sensing.

Discussion and conclusions

Despite the recognized diversity of mountain environments and regions at all scales, it is possible to recognize a common set of issues and problems for the scientist and manager who wishes to apply GIS to management and research within these regions. In addition to the data issues discussed earlier, organizational and theoretical issues are also of great importance.

Organizational issues derive particularly from the marginal locations of most mountain regions. As they are generally not investment centres, funding for many activities, including the use of GIS, may be seriously restricted. Furthermore, political sensitivities may lead to restrictions on access to data. Accessibility in mountains is not only a political, but also a physical issue, relating to both the collection and the transfer of information in harsh, rugged environments. Given the common limitations of physical accessibility in mountain areas, decentralized approaches to the implementation of GIS, as described by Pulsford and Ferrier (1994), may be particularly applicable.

Perhaps a more important issue lies in whether or not current GIS technology provides an appropriate methodological framework in which to develop models about mountain environments. As mentioned earlier, the two-dimensional nature of virtually all current GIS places considerable constraints on the latitude within which one can develop models. This is perhaps typified by the significant errors that typically derive from the use of standard GIS functions, such as the calculation of area, slope, and aspect. This is not surprising when one considers the distortion necessary to transform mountain space into a planar projection. The flattening of the mountain environment creates further problems which, as noted above, lead to significant difficulties in creating land-cover maps from remotely-sensed imagery. For instance, critical environments with rare species, such as cliffs, may be effectively invisible in current GIS because their horizontal area is so minute. Equally, it is almost impossible to capture the micro-level diversity in the spatial organization and arrangements of landscape features without the most detailed of surveys; although information about the optimum scale of analysis for specific purposes is increasingly becoming available (e.g. van Westen, 1994).

Turning to issues relating to data processing, we should note that all data processing algorithms have an implicit theory content. Data processing techniques, such as overlay, buffering, and image ratioing, contain our ideas about how mountain systems work. The important question is; how well does today's GIS toolbox reflect our knowledge about the behaviour of mountain processes? Most GIS have an array of spatial processing functions which are combined in various ways to build a process model. The sceptic would argue that a GIS approach encourages us to seek an order and a simplicity which do not exist. Take, for example, the problem of migrating mammals and the spatial analysis of distance between habitats. In this case, a mountain area is best considered as an island with impassable barriers at high and, often, low elevations. Within this territory, animals will move depending upon the availability of food, cover, and other factors so that the distance function is a subtle, non-linear combination of the spatial arrangement of habitats. The algorithms available in most current GIS may be unable to treat such complex situations satisfactorily; at best it may be possible to make broad generalizations from satellite imagery and fieldwork.

This leaves us with a fundamental dilemma: should we use present technology and data to develop a model which we know from the start will be inherently flawed, or should we seek an improved solution? We would argue that the challenge for mountain scientists is to take the latter course of action and to use existing GIS technology, explicitly recognizing its limitations, with the intention of improving both the contents of the toolbox and the methodological framework for their application. Nevertheless, there are undoubtedly some instances in which the resources required to obtain the hardware, software, and data necessary to create a GIS and then use it far outweigh the potential benefits for research or decision-making.

Finally, we should recognize another potentially serious problem: that the increased use of GIS may lead to scientists and policy-makers spending more time in the office and computer laboratory rather than doing research and working in the field. This is a situation which could isolate both policy and research and lead to a negative view of a technology which, as shown by the chapters in this volume, has much to offer for the understanding and management of mountain environments. Instead, we would promote the use of GIS as a means for integrating the understanding of natural and social scientists, the technological expertise of computer scientists, and the practical experience of managers and policy makers. As such, the value of GIS for mountain people and their environments can only grow.

References

Alegre, P., 1982, *Una aplicació del programa M.A.P. a Catalunya*, Barcelona: Departament de Geografia, Universitat Autònoma de Barcelona.

Ashdown, M. and Schaller, J., 1990, *Geographic information systems and their application in MAB-projects, ecosystem research and environmental monitoring*, MAB-Mitteilungen 34, Bonn: German National Committee for the UNESCO Man and the Biosphere (MAB) Programme.

Barry, R.G., 1992a, Climate change in the mountains, in Stone, P.B. (Ed.) *The state of the world's mountains*, London: Zed Books, 359–80.

Barry, R.G., 1992b, *Mountain weather and climate*, 2nd edn, London: Routledge.

Beissmann, H., 1989, *Plausibilitätsanalysen mit Hilfe eines EDV-gestützten thema-kartographischen Informationssystems*, Berichte und Informationen 14, Vienna: Institut für Kartographie, Österreichische Akademie der Wissenschaften.

Binz, H.R. and Wildi, O., 1986, Szenarien, in Wildi, O. and Ewald, K. (Eds) *Der Naturraum und dessen Nutzung im alpinen Tourismusgebiet von Davos*, Report 289, Birmensdorf: Swiss Federal Institute of Forestry Research, 275–314.

Bird, A.C. and Taylor, J.C., 1990, The role of GIS in monitoring landscape change in the National Parks of England and Wales, in Proceedings of 2nd National Conference of Canadian Institute of Surveying and Mapping, *GIS for the 1990s*, Ottawa, 647–56.

Brown, S.J., Schreier, H.E., Woods, G. and Hall, S., 1994, A GIS analysis of forestry/caribou conflicts in the transboundary region of Mount Revelstoke and Glacier National Parks, Canada, in this volume, Chapter 12.

Brzeziecki, B., Kienast, F. and Wildi, O., 1994, Potential impacts of a changing climate on the vegetation cover of Switzerland: a simulation experiment using GIS technology, in this volume, Chapter 14.

Bucek, A. and Lacina, J., 1981, The use of biogeographical differentiation in landscape protection and design, *Sbornite* (CSGS, Praha), **86**, 1, 4–50.

Burkmar, R.J., Petch, J.R., Basden, A. and Yipp, J., 1992, Integrating GIS and knowledge-based systems for landscape management, in *Proceedings, Conference of the Association of Geographic Information*, Birmingham, 1.28.1–5.

Butler, D.R., Walsh, S.J. and Malanson, G.P., 1990, GIS applications to the indirect effects of forest fires in mountainous terrain, in *Fire and environment: ecological and cultural perspectives*, General Technical Report SE-69, Washington, DC: US Department of Agriculture, Forest Service, 202–211.

Carver, S., Heywood, D.I. and Cornelius, S., 1993, 'Evaluating field-based GIS for environmental characterization', presentation at 2nd International Conference on Integrating Geographic Information Systems and Environmental Modelling, Breckenridge, Colorado, September.

Claudin, J., Sourp, E. and Puydarrieux, P., 1993, Constitution d'un système d'information géographique pour la gestion d'un espace naturel: l'expérience du parc national du Mercantour, in *Environmental Information Systems*, Le Bourget-du-Lac: International Centre for Alpine Environments, 31–42.

Culbertson, K., Hershberger, B., Jackson, S., Mullen, S. and Olson, H., 1994, Geographic information systems as a tool for regional planning in mountain regions: case studies from Canada, Brazil, Japan, and the USA, in this volume, Chapter 6.

d'Oleire-Oltmanns, W., 1987, MAB-Projekt 6 Habitatbewertung und potentielle Verbreitung von Tierarten unter touristicher Einfluss, *Verhandlungen der Gesellschaft für Oekologie* (Graz, 1985), **15**, 48–56.

d'Oleire-Oltmanns, W., 1990, The interaction of patchiness, land cover type and animal distribution: an evolution in time and space, in *Proceedings, Resource Technology 90*, Washington, DC, 369–75.

d'Oleire-Oltmanns, W. and Franz, H.P., 1991, Das zoologische Informationssystem (ZOOLIS) der Nationalparkverwaltung Berchtesgaden, *Verhandlungen der Gesellschaft für Oekologie* (Freising-Weihenstephan, 1990), **20**, 685–92.

d'Oleire-Oltmanns, W., Franz, H.P. and Schuster, A., 1991, Die Anwendung des Oekosystemforschung für die Analyse der räumlichen Habitatverteilung von Tierarten, *Verhandlungen der Gesellschaft für Oekologie* (Osnabrück, 1989), **19**, 3, 619–27.

Downey, I.D., Heywood, D.I. and Petch, J.R., 1990, GIS for landscape management: Zdarske Vrchy, Czechoslovakia, in *GIS for the 1990s*, Ottawa, 1528–38.

Downey, I.D., Heywood, D.I. and Petch, J.R., 1991, Landscape ecology as an operational framework for environmental GIS, in Jackson, M.C. (Ed.) *Systems thinking in Europe*, New York: Plenum, 151–8.

Forman, R.T.T. and Gordon, M., 1986, *Landscape ecology*, New York: Wiley.

Gerrard, A.J., 1990, *Mountain environments*, London: Belhaven.

Grötzbach, E.F., 1988, High mountains as human habitat, in Allan, N.J.R., Knapp, G.W. and Stadel, C. (Eds) *Human impact on mountains*, Totowa, New Jersey: Rowman and Littlefield, 24–35.

Haefner, H. and Hugentobler, F., 1985, *Assessment and monitoring of abandoned agricultural land in the Swiss Alps – Methods and Examples*, Remote Sensing Series 9, Zürich: Department of Geography, University of Zürich.

Haefner, H. and Hugentobler, F., 1988, Monitoring of natural resources for environmental and regional planning in the southern Swiss Alps, in *Proceedings, 8th EARSeL Symposium*, Capri, 368–77.

Haefner, H., Gresch, P., Hugentobler, F. and Marti, S., 1991, Landschaftswandel und Landschaftsschutz: Das Beispiel des Malcantone, Kanton Tessin, *Regio Basiliensis*, **32**, 2, 125–35.

Halpin, P., 1994, A GIS analysis of the potential impacts of climate change on mountain ecosystems and protected areas, in this volume, Chapter 15.

Haslett, J.R., 1990, Geographic information systems: a new approach to habitat definition and the study of distributions, *Trends in Ecology and Evolution*, **5**, 214–18.

Haslett, J.R., 1994, Community structure and the fractal dimensions of mountain habitats, *Journal of Theoretical Biology*, in press.

Hewitt, K., 1992, Mountain hazards, *GeoJournal*, **27**, 47–60.

Ives, J.D., 1992, Preface, in Stone, P.B. (Ed.) *The state of the world's mountains*, London: Zed Books, xiii–xvi.

Ives, J.D. and Messerli, B., 1989, *The Himalayan dilemma: Reconciling development and conservation*, London: Routledge.

Jordan, G., 1994, GIS modelling and model validation of deforestation risks in Sagarmatha National Park, Nepal, in this volume, Chapter 13.

Koshkariov, A., Krasovskaia, T.M. and Tikunov, V.S., 1994, Towards resolving the problems of regional development in the mountains of the Commonwealth of Independent States using Geographic Information Systems, in this volume, Chapter 5.

McKendry, J.E. and Eastman, J.R., 1990, Applications in forestry – review paper, in McKendry, J.E., *et al.* (Eds) *Applications in forestry*, Explorations in geographic information systems technology, Vol. 2, Worcester, Mass.: IDRISI Project, Clark University.

Michal, I., 1992, *Ecological stability*, Brno: Veronica.

Miller, D.R., Morrice, J.G., Horne, P.L. and Aspinall, R.J., 1994, Use of GIS for analysis of scenery in the Cairngorm mountains of Scotland, in this volume, Chapter 7.

Paracchini, M.L. and Folving, S., 1994, Land use classification and regional planning in Val Malenco (Italian Alps): a study on the integration of remotely-sensed data and digital terrain models for thematic mapping, in this volume, Chapter 4.

Parker, C.R. *et al.*, 1990, Natural resources management and research in Great Smoky Mountains National Park, in *Proceedings, Resource Technology '90*, 2nd International Symposium on Advanced Technology in Natural Resource Management, Falls Church: American Society for Photogrammetry and Remote Sensing, 254–64.

Pauknerova, P., Heywood, D.I., Brokes, P. and Petch, J.R., 1992, GIS and remote sensing team-up in Bohemia-Moravian Highlands study, *GIS Europe*, **1**, 4, 17–21.

Petch, J.R., and Kolejka, J., 1993, The tradition of landscape ecology in Czechoslovakia, in Haines-Young, R., Green, D.R. and Cousins, S.H. (Eds) *Landscape ecology and GIS*, London: Taylor and Francis, 39–56.

Pulsford, I., and Ferrier, S., 1994, The application of GIS by the National Parks and Wildlife Service of New South Wales, Australia to conservation in mountain environments, in this volume, Chapter 11.

Quarrie, J. (Ed.) 1992, *Earth Summit '92*, London: Regency Press Corporation.

Rodcay, G., 1991, GPS/GIS project benefits biodiversity, *GIS World* (September, 1991), 52–3.

Schaller, J., 1994, GIS and ecosystem models as tools for watershed management and ecological balancing in high mountain areas: the example of ecosystem research in the Berchtesgaden area, Germany, in this volume, Chapter 3.

Schmidt, M. and Schreier, H., 1991, Quantitative GIS analysis of the forest resources in a mountain watershed in Nepal, in *GIS-91, Applications in a changing world*, Vancouver: Forestry Canada, 227–31.

Schreier, H., Shah, P.B. and Kennedy, G., 1990, Evaluating mountain watersheds in Nepal using micro-GIS, *Mountain Research and Development*, **10**, 151–9.

Schreier, H., Brown, S., Schmidt, M., Shah, P., Sthrestha, B., Nagarmi, G. and Wymann, S., 1993, Gaining forests but losing ground: A GIS evaluation in a Himalayan watershed, *Environmental Management*, **17**, 8, in press.

Schuster, A., 1990, Ornithologische Forschung unter Anwendung eines geographischen Informationssystems, *Salzburger Geographische Materialen*, **15**, 115–23.

Schweinfurth, U., 1992, Mapping moutains: vegetation in the Himalaya, *GeoJournal*, **27**, 73–83.

Simonett, O., 1993, *Geographic information systems for environment and development*, Zurich: Geographisches Institut, Universität Zürich-Irchel.

Steiner, D. and Zamani, F., 1984, *Datenbank MAB-Grindelwald*, Fachbeitrag zum schweizerischen MAB Programm 21, Bern: Bundesamt für Umweltschutz.

Stocks, A.M. and D.I. Heywood, 1994, Terrain modelling for mountains, in this volume, Chapter 2.

Thompson, W.A., Van Kooten, G.C., Vertinsky, I., Brown, S. and Schreier, H., 1993, A preliminary economic model for evaluation of forest management in a geo-referenced framework, in *Proceedings, 7th International Symposium on Geographic Information Systems in Forestry, Environment and Natural Resource Management*, Vancouver: Forestry Canada and Polaris Learning, pp. 209–19.

Trafas, K. (Ed.) 1985, *Atlas of Tatra National Park*, Zakopane and Cracow: Tatra National Park and Cracow Section of the Polish association of Earth Sciences.

van Wagtendonk, J.W., 1990, GIS applications in fire management and research, in *Fire and environment: ecological and cultural perspectives*, General Technical Report SE-69, Washington DC: US Department of Agriculture, Forest Service, 212–14.

van Westen, C.J., 1994, GIS in landslide hazard zonation: a review, with examples from the Colombian Andes, in this volume, Chapter 8.

Vasconcelos, M.J., Pereira, J.M.C. and Zeigler, B.P., 1994, Simulation of fire growth in mountain environments, in this volume, Chapter 9.

Wadge, G., 1988, The potential of GIS modelling of gravity flows and slope instabilities, *International Journal of Geographical Information Systems*, **2**, 143–52.

Walsh, S.J., Cooper, J.W., Von Essen, I.E. and Gallager, K.R., 1990, Image enhancement of Landsat Thematic Mapper data and GIS data integration for evaluation of resource characteristics, *Photogrammetric Engineering and Remote Sensing*, **56**, 1135–41.

Walsh, S.J., Butler, D.R., Brown, D.G. and Bian, L., 1994, Form and pattern in the alpine environment: an integrated approach to spatial analysis and modelling in Glacier National Park, USA, in this volume, Chapter 10.

Weibel, R., and Heller, M., 1991, Digital terrain modelling, in Maguire, D.J., Goodchild, M.F. and Rhind, D.W. (Eds) *Geographic Information Systems*, London: Longmans, 269–97.

Willison, J.H.M. *et al.*, 1992, *Science and management of protected areas*, Amsterdam: Elsevier Science Publishers.

Yozon, P., Jones, R. and Fox, J., 1991, GIS for assessing habitat and estimating population of red pandas, Langtang National Park, in Schreier, H., Brown, S. and O'Riley, P. (Eds) *Proceedings, 5th International Symposium on Geographic Information Systems in Forestry, Environment and Natural Resources*, Vancouver: Forestry Canada and Reid Collins Ltd, 233–44.

2

Terrain modelling for mountains

Andrew M. Stocks and D. Ian Heywood

Introduction

Mountains may be defined as dynamic systems in which both the extent and variability in relief are key controlling elements. Altitude, aspect, and slope strongly influence both the human and the physical characteristics of mountain ecosystems, such as the distribution of agriculture, the location of tourist facilities, the type of forestry, micro- and local climates, and the extent of mass movement. A model of relief is therefore an essential component of a mountain geographic information system (GIS). At present, the most powerful method of representing relief is to construct a mathematical model of the earth's surface: a digital terrain model (DTM) or digital elevation model (DEM). This mathematical model can be used to derive information on height, aspect, slope angle, watersheds, radiation incidence, hill shadows, and cut-and-fill estimates which may be essential as components of a management plan or inputs to a process model.

The purpose of this chapter is to introduce DTMs to those seeking to apply GIS to mountain ecosystem management and research. An overview is provided of the background, history, and principles of DTMs. This is followed by a closer look at their use. Broad application areas are identified and reviewed, including geomorphological modelling, visualization of terrain, resource management, hazard assessment, ecological modelling, data correction, map production, and climate and hydrological modelling. Attention is then focused upon the problems and pitfalls which face the users of DTMs in mountain regions, and guidelines are provided to assist in the selection and use of appropriate data.

Digital Terrain Models: an overview

Surface modelling is a general term for describing the process of representing a physical or artificial surface by means of one or more mathematical expressions. Terrain modelling is one particular category of surface modelling which deals with the specific problems of representing the surface of the earth (Petrie and Kennie, 1987). The term digital terrain model was first used by Miller and Laflamme (1958), who defined it as the statistical representation of the continuous surface of the ground by a large number of selected points with known *x*, *y* and *z* coordinates in an arbitrary coordinate field.

DTMs allow *z* values, representing height, to be interpolated for any *x*, *y* (horizontal) coordinate in the model. They are therefore very important tools for deriving secondary spatial data sets such as slope angle, aspect, and incidence of radiation. Such data sets are invaluable for the management of mountain regions, where variation in height is as important as horizontal variation.

Three distinct stages are involved in the construction of a DTM:

1. data collection: the acquisition of terrain data in the form of points with *x*, *y*, and *z* coordinates;
2. topological structuring: the process by which the captured sample of terrain points are built into a model of the terrain; and
3. model interpolation: the process of estimating elevations in regions where no data exist (Heller, 1991).

Data collection

The data for most DTMs come from four sources: ground surveys; maps; satellites; aerial photographs (Table 2.1). A fifth method, linked to the use of global positioning systems (GPS), is now also used to provide supplementary height data to improve model representation of breakline features. These are linear landscape features characterized by abrupt changes in height, such as cliffs, ridges, and incised valleys. Information about such features is vital if a DTM is to portray reality.

Ground survey

In a ground survey, height information is collected by electronic theodolite and data loggers at sample points chosen to best characterize the form of the landscape. DTMs constructed from ground survey data are more likely to provide a robust terrain model because the surveyor can add local knowledge, thus ensuring that key terrain features such as cliff lines are recorded in detail. However, despite modern digital surveying equipment, this approach is very expensive for all but the smallest of study areas. In addition, it is also impractical in many hazardous mountain environments. Nevertheless, ground survey methods have their place in digital terrain modelling, particularly with respect to capturing breakline features to complement existing data sets.

Maps

The capture of data from existing maps requires the automatic or manual digitization of height information (usually contours and spot heights). This

Table 2.1 DTM scale and accuracy/data source selection criteria

Data source	Capture method	DTM accuracy	Coverage	Sample Application
Air survey	Stereo plotters	High	Large areas of mountain terrain	Large-scale civil construction
Satellite image	Stereo autocorrelation	Moderate	Large tracts of mountain terrain	Military strategic planning
Maps	1. Digitizing (manual or automated) 2. Raster scanning	Low–moderate	All scales but only where coverage exists. Not suitable for small areas	Low budget, low quality
GPS	Direct digital output from GPS field station	High	Small project areas. Limited to moderate terrain.	Little used. Potential for breakline capture and small project area
Ground survey	Direct digital output from theodolite field station	Very high	Very small project areas. Not possible in rough terrain	Construction site design

Based on tables by Petrie (1987) and McCullagh (1988).

method obviously suffers from the major drawback that the data are derived from a secondary source having, in most cases, already been manually interpolated from original survey data. A further problem lies in the representation of breakline features. On most paper maps, these are represented with a fair degree of artistic licence. Consider, for example, Figure 2.1 which compares the original map base for a region of mountainous terrain in the Lake District National Park, in northwestern England, with the digital product derived from it. Note how the artistic representation of cliffs on the map provides additional essential information about the character of the region. This information is lost in the digital product created from the same data source. Despite this drawback, this method is in widespread use.

Photogrammetric and remotely-sensed data

Aerial photographs and remotely-sensed data are major sources of elevation data for DTMs. Traditionally, stereoscopic methods have been applied to rectified aerial photographs to capture height information (Hoff and Ahuja, 1989). In summary, this process involves capturing height information by comparing two stereo images taken from different locations. One of the advantages of this approach is that breaklines can be more easily identified and recorded. In addition, photogrammetry, being a form of remote sensing, allows data to be captured for remote mountain areas where ground survey would be impossible.

More recently, the stereoscopic technique has been applied to stereo-matched remotely-sensed images, in a method described as stereo-auto-correlation. This method of data capture can be extremely cost-effective for large study areas which would be too costly to survey using conventional aerial photography. In addition, satellite remote sensing can be used to capture data from areas where flight is hazardous due to weather or even political reasons. However, the accuracy of this data source remains a matter for concern. For example, Fukushima (1988) cites root mean square errors of 101·6 m in mountainous areas, compared to errors of 3·4 m in areas of low relief (Rodriguez *et al.*, 1988) for data collected by the SPOT satellite. Sasowsky (1992) reports that, using SPOT-derived elevation data for a region in Alaska, error is greatest in regions exhibiting the steepest gradients and in areas containing steep bedrock knobs.

Global Positioning Systems (GPS)

GPS are a method of automatic spatial data logging originally pioneered for navigation and military use. A constellation of NAVSTAR satellites permits the user of a GPS receiver to calculate his/her position on the earth's surface to an accuracy of within one metre. Altitude can also be recorded. For example, in 1992, Canada's highest mountain, Mt Logan in the Yukon, was measured at 5959 m using GPS technology, settling a long-running debate over its exact

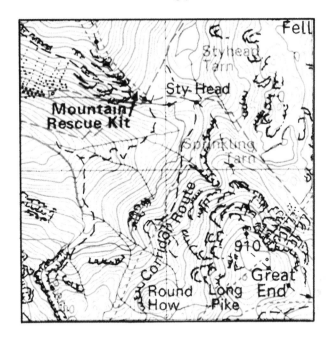

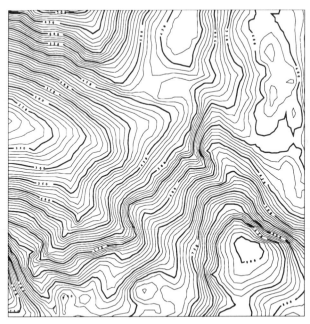

Figure 2.1 A comparison of mountainous terrain as represented on: 1. an Ordnance Survey 1:50 000 scale map (Sheet 90) of a 10 km² section of the Lake District National Park, UK; and 2. a grid-based digital terrain model (10 m contour interval, 50 m sample resolution) derived from the same base data. The map extract is Crown copyright.

GRID   TIN

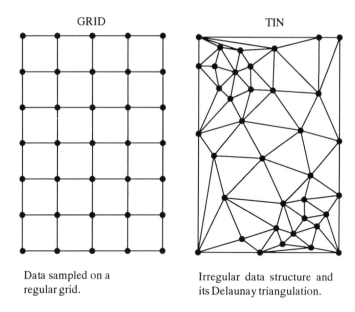

Data sampled on a   Irregular data structure and
regular grid.   its Delaunay triangulation.

Figure 2.2   A comparison of the Grid and TIN models for the construction of digital terrain models.

height. In addition, a GPS logger can be connected to a personal computer for direct data input into a GIS. For similar reasons to those associated with the use of the more traditional ground survey techniques, GPS technology is not suitable for data capture in large study areas. However, it is a cost-effective alternative to ground survey methods for the capture of breakline information. One essential prerequisite for the use of a GPS system is a clear field of view since it is necessary for the GPS receiver to fix its position from four of the NAVSTAR satellites. This situation ensures that both geographic location and height are fixed to the best accuracy. However, locking-on to four satellites in mountain terrain is often problematic when working in confined areas, such as deep valleys or mountain corries, or on forested slopes.

Model building and topological structuring

There are two main approaches for turning unstructured survey data into a usable DTM (Petrie, 1987) (Figure 2.2). In the grid method, the height data are arranged into a rectangular grid. In the triangular irregular network (TIN) method, randomly-spaced irregular point data are used to construct a triangular network of irregular size, shape, and orientation. More recently, experimental methods for the construction of DTMs based on the mathematical principles of fractal geometry have been developed for the Eastern Alps (Brivio *et al.*, 1992). These methods are not described in detail here because of the novelty of the technique. However, it is likely that this approach will be included in future GIS as understanding of its possibilities increases.

Grid-based models

The grid model is the original approach to building a DTM, first formulated and applied by Miller and Laflamme (1958). This method uses data which have been sampled or structured using a regular grid. The grid is usually rectangular or square, though non-rectangular experiments have been made (Petrie, 1986). Grid data models are sometimes referred to as lattices. Such models are relatively simple to build and make lesser demands upon computer processing resources than their TIN-based counterparts.

One useful advantage of grid-based DTM programs is that they accept data directly from digital altitude matrices, the usual format for height information captured through remote sensing and automatic photogrammetric data capture techniques. However, if an irregular (random) data set is to be used, it must be first re-sampled using a grid, so that the quality of the original data is degraded. Even with data sets already in a regular grid structure, the density of the data sampling grid must be high enough to accurately portray the most complex topographic feature present in the area being modelled. Consequently, redundancy can occur in more uniform parts of the study area. Experiments have been made to improve the simple grid model by introducing a technique of progressive sampling to increase and decrease the resolution of the sampling grid depending on relief complexity (Makarovk, 1973). However, even this process does not remedy the major drawback that grid sampling can fail to capture vital information describing abrupt changes in relief pattern. Indeed, the process of grid base sampling can actually have a generalizing effect, smoothing out such features, as shown in Figure 2.3, which portrays the same area as in Figure 2.1.

Despite these drawbacks, grid-based DTMs are in widespread use for two main reasons. First, they require less complex mathematical routines and algorithms for processing and are therefore less processing-intensive than TIN models. Second, and perhaps more importantly, raw data are very quickly and economically captured and sampled using this method.

TIN models

Triangular irregular network (TIN) models, first developed by Peuker (1978) are based on a network of triangular surfaces of irregular size, shape, and orientation. They work on randomly located height (z) values. While TIN algorithms vary, they aim to produce a set of triangles that are both as equilateral and have sides that are as short as possible. The vertices of the triangles are formed from each observed height value. The height at any point within a triangle is determined by interpolation. An important characteristic of this approach is that irregular sampling regimes can be adopted. Thus, not

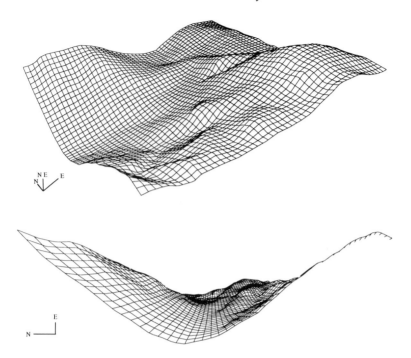

Figure 2.3   Two surface views of a 2 km × 2 km area of the Lake District National Park, UK, constructed from a DTM (50 m sample resolution). The area shown is the same as in Figure 2.1. Upper view looking northeast from a high observation point. Lower view looking east from a low observation point above the valley floor.

only can the density of data points be varied in an irregular fashion, according to the variation in relief, but breaklines can be incorporated into the sampling regime. As well as breakline information, a TIN model can hold single values for mountain peaks and depressions in the landscape.

The TIN model has three important advantages over the grid model, it can:

1. accept breakline features;
2. randomly vary the number and location of data points used in model construction;
3. reduce the volume of data required.

Naturally, the TIN model approach also has drawbacks, of which the most significant is that it is computationally intensive, due to the complex mathematical operations necessary to construct the model.

Interpolation

The user of a DTM has to extract values from the model for locations for which observed data are not available. This process, described as interpolation

(Lam, 1983), is certainly not specific to DTM applications and is a problem faced by all spatial scientists. Interpolation methods suitable for extracting information from a DTM fall into two categories: global or local.

Global methods fit a mathematical function to the entire data set in such a way that all data points are considered equal when interpolating a value at an unknown location. Such methods include techniques such as trend surface analysis. Local techniques, in contrast, assume an autocorrelation effect, so that points closer to the location for which an estimate is required determine its character to a greater degree than those further away. In general, local methods have been adopted for DTM interpolation in mountain environments. Most of these techniques are based on triangulation procedures similar to those used to construct TIN DTMs; the most common strategy for digital terrain modelling is a grid-based DTM using TIN-based local interpolation (Heller, 1991). Local interpolation methods include contouring, spatial moving averages, and more statistical techniques such as kriging.

No interpolation procedure is clearly superior to all others and appropriate for all applications (Lam, 1983). Each approach has its compromises. For example, global techniques may be more appropriate for developing an understanding of the effects of terrain on the general climatic patterns of a region, whereas local methods would be much more suitable in generating height, slope, and aspect information for inclusion in a model for agricultural development or slope stability.

Ideally, through interpolation, a DTM should be able to characterize all major elements of a landscape (e.g. valleys, ridges, summits) using a minimum number of data values. In areas of varied terrain (i.e. containing both flat and graded relief), higher-resolution sampling should be done to control the most highly variable areas, whilst fewer points are necessary to describe less varied areas. Consequently, a grid structure would be less suitable, as there would be a redundancy of data in the less varied areas in order to maintain the detail of the areas of most varied relief. However, we would suggest that, because the terrain of mountainous areas is often very varied, a simple grid structure at a suitable resolution might be very effective, and at the same time not need any complex algorithm for interpolation. Nevertheless, in a situation where all other factors are equal (in terms of available software and processing hardware resources), the character of the terrain and the requirements of the application should determine the method of data sampling and thus in turn lead to the use of the most efficient model-building algorithm for the given sampling structure.

DTMs for mountain environments

The use of DTMs for management and research in mountain environments mirrors the general trends of the use of GIS within these environments

Table 2.2   Uses of DTMs in mountain environments

Application	Example
Geomorphology	Changes in surface shape and volume of glaciers (Reinhardt and Rentsch, 1986) Concavity/convexity slope index (Parker *et al.*, 1990) Mapping earthquake epicentres (Carr *et al.*, 1990)
Visualization	Combining DTMs with digital ortho or satellite imagery to produce illuminated or shaded view models (Gillessen, 1986) Shaded relief maps (Judd, 1986) (UNEP/ICIMOD, 1990) Visualization of project impacts (UNEP/ICIMOD, 1990)
Resource management	Resource potential mapping (Hart *et al.*, 1985; Schmidt, 1991) Development planning (Haefner and Hugentobler, 1988)
Hazard assessment	Monitoring the effects of forest fires in mountain terrain (Butler *et al.*, 1990) Avalanche control (Walsh *et al.*, 1990a) Mass movement (Wadge *et al.*, 1991)
Ecological modelling	Habitat assessment (Haslett, 1990) Soil fertility (Wymann, 1991) Treeline migration (Walsh and Kelly, 1991) Vegetation mapping (Dutton, 1989)
Data correction	Geometrical correction of remotely-sensed imagery (Hill and Kohl, 1988; Brivio *et al.*, 1988; Haefner and Hugentobler, 1988)
Climate modelling	Defining microclimates (Schreier *et al.*, 1990) Modelling in alpine environments (Strobl, 1992) Improve land cover classification (Cibula and Nyquist, 1987)
Hydrological modelling	Regional discharge patterns (Turner *et al.*, 1990) Identification of drainage network (Marks *et al.*, 1984)

(Heywood *et al.*, 1994). These application areas are summarized in Table 2.2. Three of these application areas: geomorphological mapping and hazard assessment; geometric correction of remotely-sensed data, and geoecological modelling, are explored in more detail below. One of the problems associated with the production of such a summary is that it reflects current usage and the natural bias associated with finding source material. In particular, the summary has primarily been drawn from sources published in English. This is a reflection on the ability of the authors rather than the availability of material. A further problem in finding source materials is that much of the work on DTMs in mountain environments is unpublished.

Geomorphological mapping and hazard assessment

DTMs have many applications in the study of the geomorphology of mountain environments ranging from the simple mapping of features to process modelling (Wadge, 1988). For example, Carr *et al.* (1990) used data derived from a DTM of the Yucca Mountains in Nevada, USA, to help improve the mapping of the locations of earthquake epicentres. This work illustrates how the use of a DTM has allowed the location of earthquake epicentres, calculated from first principles, to be adjusted based on the location of topographic features identified from the DTM. The study is of particular importance because the Yucca Mountains have been designated as a repository for the storage of high-level nuclear waste.

The use of DTMs to help in modelling the dynamic nature of mountain ecosystems is characterized by the work of Walsh and colleagues (1990a; 1994) who have been involved in producing a probability map of where and on what terrain avalanches are likely to occur in the Rocky Mountains of Glacier National Park, Montana, USA. They use a multi-criteria approach which weights the importance of different data layers within the GIS relative to their importance in the avalanche formation process. Several of these data layers, including height, aspect and slope angle, have been derived from a DTM of the region. The study is a clear example of the use of a DTM within a GIS framework to provide valuable information for modelling physical processes. However, as the authors point out, the model is constrained by the data available for its construction. They consider their approach suitable for the broad zonation of areas susceptible to the avalanche hazard, rather than the identification of individual avalanche paths. Such detailed work would require data of a higher resolution than are currently available. Continued field research is therefore essential to validate the results produced by the model.

Geometric correction of remotely-sensed data

Remotely-sensed imagery (aerial photographs and satellite data) is a major source of data for environmental GIS applications. These sources of data have, perhaps, a greater value in mountain environments, where traditional ground-based surveying techniques are restricted by the nature of the terrain. However, most remotely-sensed images are analysed assuming flat terrain (Brivio *et al.*, 1992); a trait which, by definition, mountain ecosystems do not possess. Geometric distortion arises when either an aerial photograph or a satellite image is used to record a mountain region. In mountainous areas, the displacement between observed and true map locations of ground features has been estimated as ± nine pixels, i.e. ± 270 m for pixels with a resolution of 30 m (e.g. Landsat Thematic Mapper) (Hill and Kohl, 1988). Therefore, dealing with geometric distortions due to high variability in relief is a major factor in making sensible use of remotely-sensed data in mountain regions.

DTMs have been used to assist in the correction of both aerial photography and satellite data. Haefner and Hugentobler (1988) provide an excellent summary of the steps involved in using DTMs to correct the topographic distortion of aerial photographs. The first stage uses the DTM to help select control points for the geo-rectification of the image. This involves a projective transformation in which a model of geometric deformation is constructed by bivariate polynomial methods and the relief displacement correction is applied to all pixels on a line-by-line sequence, following the swath lines of the original image. This procedure is estimated to improve the displacement error to within one pixel. Hill *et al.* (1988) used this approach to develop a land-use inventory for the mountainous areas of the Ardeche, in France.

Other methods of using DTMs to correct satellite image data in areas of high relief include the procedure known as shadow matching (Brivio *et al.*, 1988), in which shadows identified on the satellite image are matched against shadows identified by computer simulation. The rectification procedure has also been improved by using DTM derivatives, such as slope angle and aspect, to improve classification methods (Walsh, 1990b).

Geoecological modelling

Since the physical form of mountain environments is an essential component of all aspects of geoecological modelling, DTMs are obligatory for developing geoecological models to assist in resource assessment and planning. Numerous authors (e.g. Haefner and Hugentobler, 1988; Schreier *et al.*, 1990; Parker *et al.*, 1990) have used DTMs as a key component of multi-criteria modelling. Two particular studies stand out. Walsh *et al.* (1990b) show how a DTM derived from a 1:24 000 topographic map was used to derive slope and aspect information to assist in research on treeline migration and the variability of terrain in Glacier National Park, Montana, USA. Schreier *et al.* (1990) used a DTM to derive a simple geoclimatic model with four microclimatic zones for the study of soil fertility in relation to a proposed programme of tree planting in the Jhiku Khola watershed in the Kabhre District of Nepal. In this case, the criteria used to query the DTM were derived from field investigation which revealed that virtually all the slopes below 30° are dominated by agriculture. A natural altitudinal boundary, marked by a change in the native tree species, exists at 1200 m, and the exposure of south- and north-facing slopes has a marked effect on temperature change during the dry season. A reverse process, in which a climate model is used in conjunction with a DTM, has been used to help to improve land-cover classifications in mountain regions by controlling for climatic effects (Cibula and Nyquist, 1987).

Problems and issues

Ebner (1987) argues that a DTM for use in a mountain environment should be of high fidelity, i.e. information other than height reference points should be recorded. This information should include breaklines, skeleton lines, and characteristic points. The basic problem with DTMs at typical resolutions is that they do not reveal such features. The solution to this problem lies in the use of a DTM of the appropriate detail and resolution.

The accuracy of a DTM depends on the spatial density and accuracy of the sampled data points. Most DTMs have been captured at scales from 1:1000 to 1:25 000, and are usually derived from paper contour maps. Such DTMs inherit many of the problems associated with this source of information and, as a rule of thumb, have a standard error that is usually half the contour interval (Wadge, 1988). The problem is that the integrity of the sampling framework influences the quality of derived data. For example, Walsh *et al.* (1987) found errors in slope angle and slope aspect, summarized for 100 m and 200 m grids respectively, were consistently out by 50 per cent when checked against field data.

The level of precision required in a DTM depends on its purpose (Walsh *et al.*, 1987). For example, Pearson *et al.* (1991) note that DTMs generated from 1:1000 topographic maps with 16 m contour intervals were insufficient to depict most of the landslide source scars in the mountains of western central Cyprus. In addition, they suggest that, in this case, the DTM data derived from a 1:1000 scale topographic map would be unsuitable for the physical modelling of landslides. Wadge (1988) takes the argument further by looking at the three specific components associated with the modelling of slope stability. He notes that most DTMs are inappropriate for determining the dynamics of slope stability where the value of every pixel is crucial to the model. Similarly, he points out that dynamic slope flow models derived from DTM data of the wrong resolution may lead to the scale of the problem being magnified, and material sent down the incorrect path. In contrast, slope energy balance models will be less sensitive to individual pixel values, so that it is quite plausible for DTMs derived from topographic maps at scales greater than 1:1000 to provide an overview suitable for the development of zonal mapping. These conclusions seem to agree with the study of avalanche tracks by Walsh *et al.* (1990a). Many of these problems are enhanced when three-dimensional (3D) perspective displays are used; for example, hill shading or errors in slope mapping will be highlighted because of gradient anomalies, rivers will appear to flow uphill, and false ridges, passes, and cols may be identified.

The only way to solve most of the problems and pitfalls identified above is to validate the DTM through a comprehensive programme of comparison with field observations. Best-fit methods such as the root-mean-square-error approach can be used to enhance this approach; for a comprehensive review of these techniques, see Willmott (1984). More complex statistical methods such as comparative analysis of semivariograms or frequency spectra (Weibel and Heller, 1991) may also be used. However, for most applications neither the time nor the available statistical expertise permit such an approach. Perhaps the most pertinent advice is provided by Walsh *et al.* (1987), who state that a user needs to evaluate the appropriateness of the DTM products for specific applications and to ensure that the spatial integrity of the DTM is always below the smallest terrain feature for which information is required.

Conclusions

At the present time, DTMs are an essential component of GIS for modelling and managing mountain environments, as they represent the major techniques for including the 3D nature of mountains in GIS (Heywood *et al.*, 1994). Used with care, DTMs provide a valuable tool for developing insights into the physical and human processes that shape mountain regions, as illustrated by the various case studies reviewed in the chapter. Ideally, the mountain scientist requires a truly 3D GIS. However, technical and methodological constraints presently restrict wide-scale use of such an approach to all but the basic visualization of information. In addition, it should be remembered that DTMs are only approximate representations of reality, requiring continual verification and improvement through field research. New techniques such as fractal geometry may provide us with a better model in the future but, at present, we must rely on the available data and use our own judgment to assess the validity of decisions made using such models.

References

Brivio, P.A., Della-Ventura, A., Rampini, A. and Schettini, R., 1988, Shadow structural description as a tool for automatic registration, in *Alpine and Mediterranean areas*, Proceedings, 8th EARSeL Symposium, Capri, 392–8.

Brivio, P.A. *et al.*, 1992.

Butler, D.R., Walsh, S.J. and Malanson, G.P., 1990, GIS applications to the indirect effects of forest fires in mountainous terrain, in *Fire and the environment: ecological perspectives*. Knoxville, Tennessee, March 20–24, pp. 202–11.

Carr, J., Keryl, F., Aymard, W. and Taylor, J., 1990, Application of a raster based GIS technique to temporal monitoring at Yucca Mountain, Nevada, USA, in Proceedings of *GIS National Conference, Ottawa, Challenge for the 1990s*.

Cibula, W.G. and Nyquist, M.O., 1987, Use of topographic and climatological models in a geographical database to improve Landsat MSS classification for Olympic National Park, *Photogrammetric Engineering and Remote Sensing*, **53**, 67–75.

Dutton, L., 1989, Vegetation mapping in the Northern Rocky Mountains, in General Report INT-57, US Department of Agriculture, Forest Service, 97–104.

Ebner, H., 1987, Digital terrain models for high mountains, *Mountain Research and Development*, **7**, 353–6.

Fukushima, Y., 1988, Generation of DTMs using SPOT Image near Mt Fuji by Digital

Image Correlation, *International Archives of Photogrammetry and Remote Sensing*, **27**, B3, Commission III, 225–34.

Haefner, H. and Hugentobler, F., 1988, Monitoring of natural resources for environmental and regional planning in the Southern Swiss Alps, in *Alpine and Mediterranean areas*, Proceedings, 8th EARSeL Symposium, Capri, 368–78.

Hart, J.A., Wherry, D.B. and Bain, S., 1985, An operational GIS for Flathead National Forest, in *Proceedings, Autocarto 7*, Washington, DC: American Society of Photogrammetry and American Congress of Surveying and Mapping, 244–53.

Haslett, J.R., 1990, Geographic Information Systems: A new approach to habitat definition and the study of distributions, *Trends in Ecology and Evolution*, **5**, 214–18.

Heller, M. and Weibel, R., 1991, Digital terrain modelling in Maguire, D.J., Goodchild, M.F. and Rhind, D.W. (Ed.) *Geographical information systems Vol. 1 Principles*, Longman. Ch. 19, pp. 251–67.

Heywood, D.I., Price, M.F. and Petch, J.R., 1994, Mountain regions and geographic information systems, in this volume, Chapter 1.

Hill, J. and Kohl, H.G., 1988, Geometric registration of multi-temporal thematic mapper data over mountainous areas by use of a low resolution digital elevation model, in *Alpine and Mediterranean areas*, Proceedings, 8th EARSeL Symposium, Capri, 323–35.

Hoff, W. and Ahuja, N., 1989, Surfaces from stereo: integrating feature matching, disparity estimation, and contour detection, *IEEE Transactions on Pattern Analysis and Machine Intelligence*, **11**, 121–35.

Judd, D.D., 1986, Processing techniques for the production of an experimental computer generated shaded relief map, *The American Cartographer*, **13**, 72–9.

Lam, N.S., 1983, Spatial interpolation methods: A review, *The American Cartographer*. Vol. 10, No. 2, pp. 129–49.

Makarovk, B., 1973, Progressive sampling methods for digital elevation models, *ITC Journal*, **3**, 397–416.

Marks, D., Dozier, J. and Frew, J., 1984, Automated basin delineation from digital elevation data, *Geo-Processing*, **2**, 299–311.

McCullagh, M.J., 1988, Terrain and Surface Modelling Systems: Theory and Practice, *Photogrammetric Record*, **12**, 747–79.

Miller, C.L. and Laflamme, R.A., 1958, The digital terrain model – theory and application, *Photogrammetric Engineering*, **24**, 422–33.

Parker, R.C., Langdon, K. Carter, J.R., Noduin, S.C. and Barett, H., 1990, Natural Resources Management and research in Great Smoky Mountains National Park, in *Proceedings, 'Resource Technology '90'*, 2nd International Symposium on Advanced Technology in Natural Resource Management, Falls Church: American Society for Photogrammetry and Remote Sensing, 254–64.

Pearson, E.J., Wadge, G. and Wislocki, A.P., 1991, Mapping natural hazards with spatial modelling systems, presented at *EGIS '91*. Brussels, Belgium April 2–5, 1991.

Petrie, G., 1986, Data interpolation and contouring methods, in *Proceedings, Terrain modelling in civil engineering*, Vol. 1, Glasgow.

Petrie, G. and Kennie, T.J.M., 1987, An introduction to terrain modelling applications and terminology, in *Proceedings, Terrain modelling in civil engineering*, Vol. 1, September 1–3 University of Glasgow, Scotland.

Peuker, T.K., 1978, Data structures for digital terrain models: discussion and comparison, in *Harvard papers on GIS*, **5**, 1–5.

Reinhardt, W. and Rentsch, H., 1986, Determination of changes and elevation of glaciers using digital elevation models for the Vernagtferner, Oetzal Alps, Austria, *Annals of Glaciology*, 151–5.

Rodriguez, V.P., Gigord, A.C., De Gawjac, P., Monier, P. and Begrug, G., 1988,

Evaluation of the Stereoscopic Accuracy of SPOT Satellite, *Photogrammetric Engineering and Remote Sensing*, **54**, 217.

Sasowsky, K.K., 1992, Accuracy of SPOT digital elevation model and derivatives: utility for Alaska's north slope, *Photogrammetrical Engineering and Remote Sensing*, **58**, 815–24.

Schmidt, M., 1991, An evaluation of the forest resources and forest soil fertility of a mountain watershed in Nepal using GIS techniques, in Shah, P.B. *et al.* (Eds) *Soil fertility and erosion in the Middle Mountains of Nepal* Kathmandu: Integrated Survey Section, Topographical Survey Branch, 244–52.

Schreier, H., Shah, P.B. and Kennedy, G., 1990, Evaluating mountain watersheds in Nepal using micro-GIS, *Mountain Research and Development*, **10**, 151–9.

Strobl, J., 1992, Climatological modelling for Alpine regions, in *Proceedings, 1st Tydac European SPANS User Conference*, Amsterdam.

Turner, K.A., Kolm, K.E. and Downey, J.S., 1990, Potential applications of three-dimensional geoscientific mapping and modelling systems to regional hydrological assessments at Yucca mountain, Nevada, presentation at International Symposium on 3-D Computer Graphics in Modelling Geologic Structures and simulating Geologica Processes, University of Freiburg, Germany, October.

Wadge, G., 1988, The potential of GIS modelling of gravity flows and slope instabilities, *International Journal of Geographical Information Systems*, **2**, 143–52.

Wadge, G., Wislocki, A.P. and Pearson, E.J., 1991, Spatial analysis in GIS for natural hazard assessment, in *Proceedings, First International Conference/workshop on Integrating Geographic Information Systems and Environmental Modelling*, Boulder, Colorado, September.

Walsh, S.J., Lightfoot, D.R. and Butler, D.R., 1987, Recognition and assessment of error in geographic information systems, *Photogrammetric Engineering and Remote Sensing*, **53**, 1423–30.

Walsh, S.J., Butler, D.R., Brown, D.G. and Bian, L., 1990a, Cartographic modelling of snow avalanche path location within Glacier National Park, Montana, *Photogrammetric Engineering and Remote Sensing*, **56**, 615–21.

Walsh, S.J., Cooper, W.J., Von Essen, I.E. and Gallager, K.R., 1990b, Image enhancement of Landsat Thematic Mapper data and GIS data integration for evaluation of resource characteristics, *Photogrammetric Engineering and Remote Sensing*, **56**, 1135–41.

Walsh, S.J., Butler, D.R., Brown, D.G. and Bian, L., 1994, Form and pattern in the alpine environment: an integrated approach to spatial analysis and modelling in Glacier National Park, USA, in this volume, Chapter 10.

Weibel, R. and Heller, M., 1991, Digital Terrain Modelling, in Maguire, D.J., Goodchild, M.F. and Rhind, D.W. (Eds) *Geographic Information Systems*, London: Longmans, 269–97.

Willmott, C.J., 1984, On the evaluation of model performance in physical geography, in Gaile, G.L. and Willmott, C.J. (Eds) *Spatial Statistics and Models*, Dordrecht: Reidel, 443–60.

Wymann, S., 1991, Land-use intensification and soil fertility in Agricultural land: A case study in the Dhulikel Khola Watershed, in Shah, P.B. *et al.* (Eds) *Soil fertility and erosion in the Middle Mountains of Nepal*, Kathmandu: Integrated Survey Section, Topographical Survey Branch, 253–9.

PART I

Regional Resource Inventory and Planning

3

Geographic information systems and ecosystem models as tools for watershed management and ecological balancing in high mountain areas: the example of ecosystem research in the Berchtesgaden area, Germany

Jörg Schaller

Introduction

The significance of geographic information systems (GIS) for environmental management and resource planning has increased in recent years. One main reason is the need to compare a great number of spatial data describing the relevant natural resources and their sensitivity to human activities. Since GIS can be used to couple spatial data with their attributes, and then to overlay these, it is highly efficient for such planning tasks. This paper describes some of the basic application methods using a GIS in connection with ecological model appoaches from work in Berchtesgaden Alpine and National Park, in the Alps of Bavaria, Germany. This work began as part of UNESCO's Man and the Biosphere (MAB) project area 6 (impacts of human activities on mountain and tundra ecosystems) (Kerner, Spandau, and Köppel, 1991) and continues as part of ongoing research and management in the Park.

Human influence on high mountain ecosystems

Current ecological theory, in particular ecosystem theory, is characterized by a new and better understanding of ecosystem patterns and dynamics. The traditional emphasis on relatively descriptive analyses of ecosystems and landscape has been replaced by an emphasis on process and feedback effects. A consideration of people and their socio-economic systems is also incorporated into such ecosystem research.

The resources of individual ecosystems are predominantly vertically orientated in the catchments of mountain ranges. In steep terrain, flows of surface and sub-surface water carry materials from ecosystems at higher levels to those at lower levels. Through these flows, low-lying ecosystems tend to accumulate soil material and dissolved nutrients from higher levels, thus increasing their overall productivity (Figure 3.1). Any human use of an environment results in the removal of materials (and their embodied energy) which may later not be broken down and returned to the same site. This means that processes of removal and supply of materials characteristic of human use replace the natural processes to a greater or lesser degree. Thus, one may speak of 'removal ecosystems' and 'supply ecosystems' defined by the transport of material, analogous to natural mountain ecosystems. As the typical natural relationships between production, use, and decomposition in producer-consumer-decomposer cycles are disturbed, anthropogenic imbalance comes to predominate, bringing with it instability (Haber and Schaller, 1988; Schaller, 1992).

The ecosystem pattern of a culturally-dominated landscape can be stabilized by maintaining a certain proportion of near- or semi-natural ecosystems in the landscape because these have the effect of minimizing the influence of the human-influenced neighbouring systems (Haber, 1972, 1984, 1986b). Consequently, the ecosystem types within a landscape may be ranked according to the principal land use types, conceived as compartments of regional natural landscapes. These can be ordered according to decreasing naturalness, with simultaneously increasing human influence.

Each land use type (and subtype) is assigned a list of typically associated environmental impacts. These are classified according to material or non-material impacts and the natural resources subject to the impacts: air, water, soil (substrate), biota, and landscape (here regarded as a composite resource, including its physiognomy). A second group involves an impact-effect hierarchy: e.g. impacts on the atmosphere tend to be more significant than impacts on the soil, which tend to remain locally restricted.

The model approach

The model approach developed for the German MAB-6 Project was developed with the help of the experience and results of the Austrian and Swiss MAB-6 Projects (Moser, 1987; Messerli, 1986). These were the first research projects in mountain areas to incorporate both integrated investigations of human-environment ecosystems and the definition of important impact factors and their interrelationships. The Swiss MAB-6 model (Messerli and Messerli, 1979; Messerli, 1986) (Figure 3.2) shows a schematic representation of a regional economic-ecological system, composed of three main components: a natural system with abiotic and biotic resources, a land use system, and a socio-economic system. All the individual components of such a model are interrelated, and react as a whole under external influences. An example

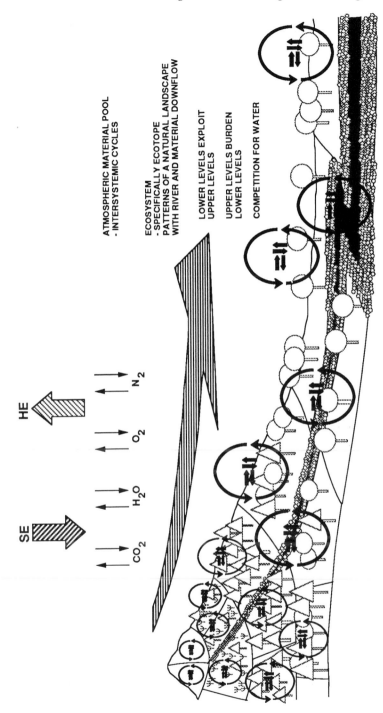

Figure 3.1 Pattern of natural ecosystems on a mountain slope. Surface runoff, streams, and rivers carry materials down to lower levels. The site-bound (intra-systemic) ecosystem cycling becomes poorer at higher levels and richer at lower levels, so that there is an inter-systemic dependency between the levels. SE = solar energy, HE = heat energy (modified from Haber, 1986b).

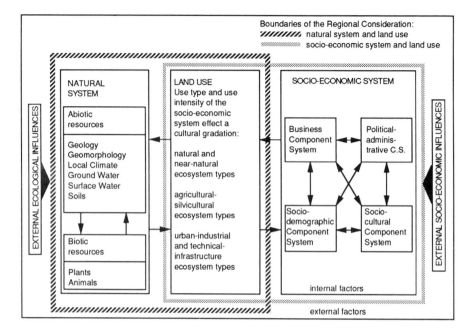

Figure 3.2   Integrated model of a regional human-environment system (modified from Messerli and Messerli, 1979).

of an external ecological influence might be the uncontrolled entry to the system of air pollutants from a distant source. An external socio-economic influence might be a political programme resulting in changed or increased agricultural and forestry use. The Swiss MAB-6 model has been applied in the Berchtesgaden MAB-6 Project at different levels of consideration determined by a number of biological and sociological specialists (Grossmann, 1983a).

The application of this theoretical approach for ecosystem management required the development of a methodology to analyse the environment with its complex ecological and socio-economic dynamic interrelationships. It was therefore necessary to translate regional system dynamics into predictive statements for different spatial and temporal scales. This complicated task was tackled by adopting the Hierarchical Systems Method proposed by Grossmann (1983b), which relies on the theories of hierarchical multilevel systems (Mesarovic *et al.*, 1971), and orderly adaptive structure (Rappaport, 1977).

An explanation of the hierarchical systems method requires a short digression (Figure 3.3). The method distinguishes two principal 'functional domains' of ecological systems:

• The process domain (or reality level), where energy and matter are processed, and biomass is produced, consumed, and decomposed. This is the everyday reality of the ecosystem, with inputs, outputs, and related processes that are readily observable, measureable, and computable.

• The regulation and control domain (or strategic and dynamic levels), where information regulating inputs, flows, storage, and outputs of energy and matter is produced and processed, resulting in the steady state of the ecosystem. In contrast to the process domain, the regulating items or agents are less readily (and sometimes unable to be) observed, measured, and/or computed.

Most ecosystem research has been devoted to the process domain, largely neglecting the regulation and control domain with the exception of some direct process-related regulation. Hence there is a prevalence of simple experimental or descriptive, rather than selective approaches. The results of such research, however, have not proved very useful in understanding system regulation and in predicting its possible behaviour. Yet it is exactly this understanding that is required in environmental planning and management in order to assess the impacts of human activities on the landscape, on ecosystems and their compartments and components, or on natural resources in general.

Landscape ecologists and ecosystem investigators have, of course, been aware of the existence and importance of the regulation and control domain. Yet it has proved almost inaccessible to experimental research because of its great complexity. Ecosystem modelling, much promoted by the rapid progress of electronic data processing, was seen as the way to a better understanding. However, the first modelling approaches failed because the complexity of the domain is not amenable to a complete holistic scientific approach as expected.

Therefore, three hierarchically ordered levels have been conceived, each mirroring the regional system of Figure 3.2. Their arrangement is hierarchical and may be represented as a three-storied pyramid (Figure 3.3). The tapering of the pyramid, causing a reduction of the dimension of each subdomain from bottom to top, symbolizes decreasing internal regulation and control powers and increasing external influences. Accordingly, the three levels have to be examined using different approaches and methods, the results of which are integrated by a special procedure called soft coupling.

The lower process level is directly and intimately connected to the process domain, or reality level. Every particular ecosystem process is thus rather strictly regulated and controlled. Data are quite easily and precisely measurable, and can therefore be obtained in abundance. Processes and interrelations are also abundant, but are mostly simple and linear; in spite of their numbers, their structure is clear, and their outcomes are predictable. Therefore, this lower level is often included in ecosystem investigations and even incorporated into GIS to produce input-output balance sheets. It also makes statistical operations possible. However, a comprehensive understanding of the operation of ecosystems cannot be gained by investigation of this level

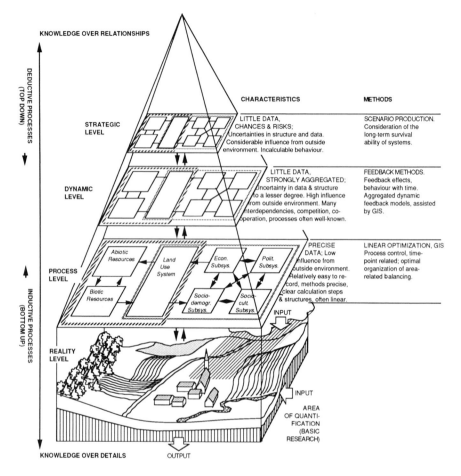

Figure 3.3   Graphic representation of the hierarchical systems method as applied to an ecosystem complex (adapted and modified from Grossmann, 1983).

alone. The perception derived from it lacks consideration of the adaptive dynamics that are essential for the survival of the ecosystem.

Operation in the dynamic level differs greatly from the rather strict regulation principles typical for the process level. It is less predictable and characterized by uncertainties in the data, in spite of a plausible structure. Internal variability of components appears to be essential for regulating irregular influences, as mentioned above. However, indicator organisms and processes, which can be derived from bottom (process) level studies, may assist in the investigation of middle-level dynamics. In short, while regulation at the bottom (process) level is concerned with the normal everyday operation of the ecosystem(s), normal but irregular events or disturbances are mastered as far as possible at the middle (dynamic) level.

The top (strategic) level of the regulation and control domain is the most difficult to explore. The external influences from the ecosystem's environment are very strong, unpredictable, and even erratic, and may endanger the survival of the system. Responses may be called strategic, but are highly uncertain and barely predictable. Not only data but also the structure is unknown. Thus, feedback approaches which are based on fixed structures are no longer appropriate, experimental approaches even less so. The investigator examining this strategic level can only work with a small amount of highly aggregated information of low accuracy and questionable value. Therefore, the dynamics of the top level are presented in scenarios and other more speculative approaches, with intrinsic subjectivity.

Application of the hierarchical systems method

Having explained the theoretical background, the problem of ecologically sound management of high mountain ecotopes affected by human actions or activities is considered again. This requires ecotope-specific predictions of consequences including normal environmental dynamics. For this purpose, the results of the investigation of the sites or ecotopes, which largely correspond to the process domain of the ecosystem(s) under consideration, have to be integrated, or at least related, to the three levels of the regulation and control domain. Whereas electronic data processing is most useful for investigating the individual levels, the integration of the results seems, at best, only partly manageable by mathematical treatment. Consequently, such integration should not be left to the computer, but must involve human judgment and even intuition. From previous experience, it is strongly recommended that a local expert who is scientifically familiar with the ecotopes and ecosystems, and in particular with their past history, carries out this integration of the results (Haber, 1990).

Geographic information system

The geographic information system (GIS) applied for ecosystem research in Berchtesgaden Alpine and National Park is used to store and process geographic information. All of the data from the study area have been stored in the form of thematic maps and related attributes using ARC/INFO software. This permits easy retrieval, overlay and presentation in map or tabular form or as a three-dimensional view for a particular purpose. Plate 1, for example, is a three-dimensional GIS output of the Berchtesgaden area. One of the most important functions of the GIS is to allow the results of data processing for intensive test areas to be transferred over the entire study area. This means the characteristics of the entire study area can be mapped in one form in one large databank which can be easily manipulated for later computer modelling.

Two types of geographic information are processed with the GIS: cartographic data, which describe the position of the mapping elements and the topological relationships between them; attribute data, which are tabulated data, identifying and describing the mapping elements to which they belong. The data are stored in tables and managed with a relational Databank Management System (DBMS) for easy access and processing (Schaller, 1987, 1988).

Initially, groups of professionals decide which variables should be represented in the databank and how to evaluate the information from the many different existing maps, showing for example geology, topography and vegetation. This is supplemented by satellite data (LANDSAT, SPOT, etc.), aerial false-colour infra-red photographs, orthophotographs, and a photogrammetric evaluation of elevation. Highly detailed data from field surveys were digitized while numeric data have been entered via terminals. Because the environment is dynamic, a time element must be considered: the year of data collection.

The GIS structure was built up as illustrated in Figure 3.4 (Schaller and Spandau, 1987). The data from the various sources are processed and combined to give firstly three main spatial data classes, from which maps can be produced:

- site data: from maps, survey results and literature, including information on rock types, soils and soil fertility, erosion, and specific site habitat information;
- topographic data: typically slope, drainage, exposure, and altitude as determined from photogrammetric measurements; and
- land use data, representing the spatial expression of human activities in the environment: agriculture, forestry, water bodies, settlements, technical infrastructure, roads etc.

The data are derived from interpretation of false-colour infra-red aerial photographs and orthophotographs. An interpretation key developed for the whole study area has been derived from information collected from the intensive test areas. A total of 196 land use types has been identified for the Berchtesgaden area.

While the mapping scale of 1:10 000 is sufficient to show most land use types, it is not so easy to show linear or point features such as streams, paths, hedgerows, or individual trees. The same problem exists for details of technical infrastructure, such as mountain cableways. These requirements necessitated the preparation of extra maps of linear and point features. Discrete mapping coordinates were applied for the latter.

The three spatial maps were overlaid to form a combined map, composed of smaller, delimited areas of common characteristics. This new file contained the 'smallest common geometry' (SCG file), the finest spatial breakdown of the study area's characteristics. Each area or polygon from the file has a specific and complete set of characteristics. This file is the one used

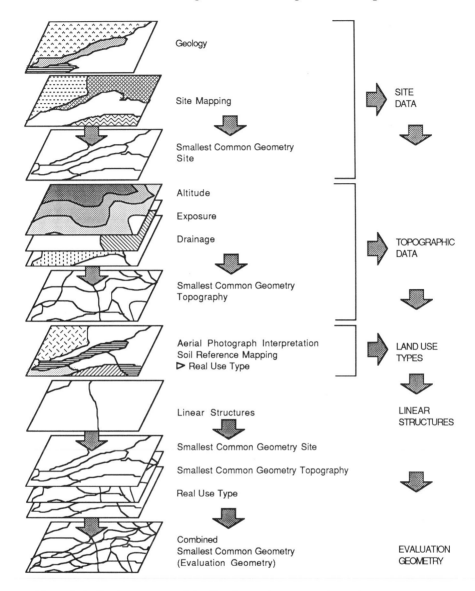

Geology

Site Mapping

SITE
DATA

Smallest Common Geometry
Site

Altitude

Exposure

Drainage

TOPOGRAPHIC
DATA

Smallest Common Geometry
Topography

Aerial Photograph Interpretation
Soil Reference Mapping
▷ Real Use Type

LAND USE
TYPES

Linear Structures

LINEAR
STRUCTURES

Smallest Common Geometry Site

Smallest Common Geometry Topography

Real Use Type

Combined
Smallest Common Geometry
(Evaluation Geometry)

EVALUATION
GEOMETRY

Figure 3.4   Basic structure of the Berchtesgaden geographic information system.

for further computer evaluation, along with the linear and point files. Mapping, statistical procedures, and modelling techniques may be applied to all the files. Specific data may be combined or aggregated as required for a particular evaluation.

Three thematic tables are prepared from all the data within the SCG-polygons to tabulate the individual characteristics or variables:

- a soil table (BODTAB) which lists soil characteristics interpreted from national mapping sources;
- a waterbody table (GEWTAB) which shows water source types (rivers, lakes, groundwater etc.) and water quality; and
- a use table (NUTZTAB) which lists all the different land use types. This includes details of various environmental factors such as vegetation type, transpiration, unproductive evaporation, root depth, sensitivity to alterations in the water budget, and sensitivity to fertilizer materials. There is also information on the ecological functions of each land use type, as well as socio-economic factors.

This data structure permits the definition of particular land uses, soil types, and waterbodies, and computer mapping of the areas where they occur. Variables can be mapped either singly or as a combination, the information being on call as required. Formal criteria are developed to evaluate the plausibility of data for each variable in the table (Schaller and Spandau, 1987). Up to sixteen points are used to describe and denote each variable. These include coding, nature and quality of data collection, and extent of occurrence.

The construction of the GIS thus makes it possible to spatially define land use types that are more or less unified, in relation to particular abiotic sites. From the initial database, modelling and representation can be achieved by altering the characteristics related to hypothetical changes. The coding system allows a whole range of changes in any one polygon to be represented by simply changing from one code number to another. It is also possible, in a modelling scenario, to develop new land use categories not already present in the study area. All these features allow the Ecological Balance Model introduced in the following section, to be run.

The Ecological Balance Model

Ecological correlations within a studied region, and effects from changes of land use over a particular area, always involve exchanges of materials and energy in space. These exchanges are in theory measurable and representable, given sufficient data and understanding of the processes in the region. An Ecological Balance Model (Kerner, 1983) has therefore been developed, to determine:

- the effects of different land uses through their interactions with neighbouring areas; and
- the ecological changes in relationships within a studied region.

The model is useful for:

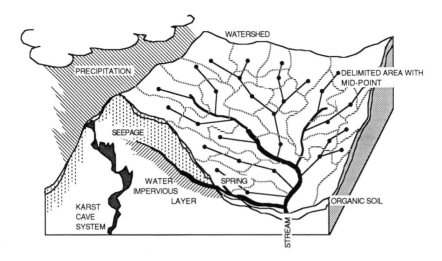

Figure 3.5 Schematic representation of water movements in the landscape, with flow 'cascading' calculated between delimited area mid-points (modified from Cejka, 1986).

- environmental analyses (research- or planning-orientated); analysis of planning alternatives (for status quo or new proposals);
- seeking to optimize the balance between ecological and economic factors;
- assessing the degree of fulfilling environmental quality standards; or
- confirming ecological and economic theories.

The model is especially applicable for determining aspects of spatial patterns and an integrated consideration of interrelated factors; and planning measures and procedures. It may be applied at individual or combined GIS SCG polygons in a study region; or functional ecological, economic or administrative units, defined as necessary.

Factors of natural water, wind and gravity transport are calculated in the balance model with the help of topographic, structural, and climatic map data. The model considers details of balance of supply by noting the locations of origin and eventual deposition of materials. This permits the determination of natural and technically-aided transport boundaries (which are considered ecological). These give functional ecological reference spaces within the studied areas. The types of materials transported depend on the nature of these areas. The model is normally applied within hydrologically-defined areas (one or more catchments), or within tourist or other activity areas (Figure 3.5) (Cejka, 1986; Haber and Schaller, 1988).

It is usually easier to describe material and energy outputs from an area. It is these components of the material and energy relationships which affect elements of other areas. These variables, stored in an OUTPUT file, are therefore very important in the model. They are also important for an evaluation of ecological sensitivity that is independent of precise spatial characteristics. Reactions of inputs and outputs in an ecosystem are expressed as a sum of all influences, by the type and degree of each influence. The extent of the influences from outside may be related to the degree of modification of a system from natural to cultural. A near-natural system is more self-regulating, while a system with increased technological influence involves greater steering from outside the natural component of the ecosystem. However, all ecosystems function in much the same way, involving transfers of physical and chemical elements, regardless of the degree of human influence. Thus, for the balance model, the discernible differences between ecosystems are considered as more quantitative than qualitative.

The material and energy relationships within different systems can be described by one variable set and quantified with the one model. The socio-economic aspects of the model relate to cost-equivalents, calculated for the matter and energy variables from the NUTZTAB table. The working of the ecological system is described with the equation:

$$\text{OUTPUTS} + \text{EMISSIONS} = \text{INPUTS} + \text{IMMISSIONS}$$
$$- \text{INTERNAL CONSUMPTION} + \text{SUPPLY CHANGES}$$

The result of a balance model run, such as the output of surface water from different land use types, can be depicted on a map. It is very easy to change land use types in the computer by changing the land use codes related to different points in time. An example might be to change the land use type 'near-natural mixed mountain forest' into the land use type 'clearcutting.' The related attributes of the NUTZTAB table, for example, transpiration, species diversity, and interception, are changed accordingly. The related attribute values for each type of ecosystem are taken for the creation of balance sheets, such as changes in species numbers and quality, soil, vegetation cover, and changes in interception. The time-related changes of land use types can also be used as input values for a second run of the ecological balance model.

An important step of the balance calculation after a land use change within a catchment is the calculation of the new surface water runoff. Figure 3.6 shows the output of a triangular irregular network (TIN) cascading calculation, with the various directions of flow and the generalized effects from the origin systems to the drainage system. As a result of the cascading procedure, the calculated soil losses and matter transportation related to surface water flows can be incorporated into the map. This calculation is needed to determine the matter and embodied energy losses from highland ecosystems to lowland ecosystems (Flacke, 1990a, b).

Dynamic feedback models

These more or less static calculations can be related back to the dynamic level of feedback models shown in Figure 3.3. The dynamic model handles

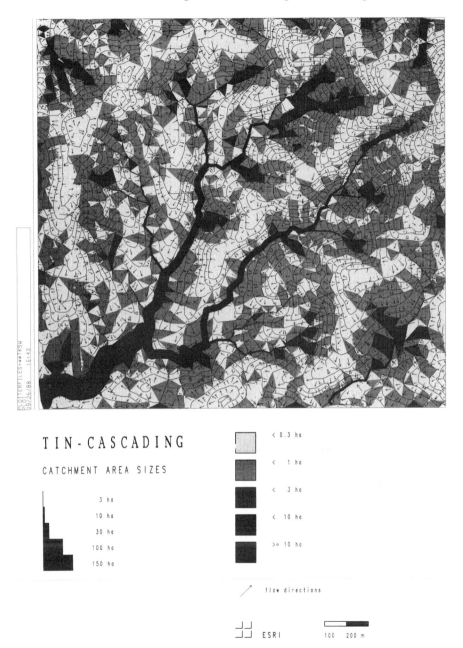

Figure 3.6 Output of a Triangular Irregular Network (TIN) cascading calculation showing flow directions and generalized effects within catchment area.

changes in land use types over time. With the definition of risks from the key variables of the GIS database, the dynamic effects can be represented in time maps or series of time maps (Grossman *et al.*, 1983). Every real or modelled situation of the landscape at a discrete point in time can be taken again as an input for the balance model to describe the ecological effects of changes in land use. This new model approach developed in the MAB-6 project in Berchtesgaden is a highly sophisticated tool for ecological modelling and balancing of matter, energy, and information flows in an environmental system.

An important example of this approach has been carried out for the study of high mountain agriculture in Berchtesgaden (Spandau and Tobias, 1990; Grossmann, 1989). From the relationships of socio-economic factors, mainly caused by externally-steered influences, such as market prices; tax reduction; state subsidy programmes or additional income from tourists, the dynamic model calculates land use changes over time caused by different land use intensities and impacts under three scenarios: trend scenario; termination of high mountain agriculture by economic constraints; and continued agricultural land use under different economic conditions. This study resulted in the definition of risk maps for abandoned areas from the GIS database, using criteria such as elevation, soil quality, distance to farm. The translation of the results of the different model runs led, in combination with the risk map, to different scenarios. These maps can be used again on the process level to run the balance model to check if a certain land use pattern has negative or positive impacts on natural resources, whether abiotic factors, such as surface water runoff or erosion, or biotic factors, such as species diversity, quality of landscape scenery related to woodland cover, and succession.

Conclusions

The support of a GIS was essential for the ecosystem research in Berchtesgaden Alpine and National Park, which set new standards for interdisciplinary research in Germany. In particular, this research has resulted in the establishment of official standards and guidelines for ecological research. Examples include procedures for environmental monitoring in National Parks and Biosphere Reserves, and new methods for applications in ecological planning (including ecological balancing techniques), and in the production of landscape management and development plans.

At a local level, the research programme resulted in the production of an environmentally-friendly traffic/public transport concept, a tourism concept considering socio-economic factors and infrastructure requirements, and a concept for mountain agriculture (Kerner *et al.*, 1991). A major study, considering several variants, was completed on the effects of proposed Winter Olympic Games in the region (Haber, 1986a). Furthermore, the administration of the National Park benefited greatly from the experience gained by using GIS. Through the project, not only did the GIS become a working instrument, it permitted the creation of a working database for the National Park, a documentation base for research, and a specialist-orientated databank which could be fully applied for practical applications in research and education. The GIS is now being used as one of the bases for the development of a master plan for the National Park.

References

Cejka, M., 1986, Methodische Aspekte einess Ökologischen Bilanzmodells im Rahmen des MAB Projektes 6 'Der Einfluß des Menschen auf Hochgebirgsökosysteme im Alpen – und Nationalpark Berchtesgaden' . . . Diplomarbeit am Lehrstuhl für Landschaftsökologie TU München-Weihenstephan.

Flacke, W., 1990a, TIN-CASCADING ARC/INFO-based drainage model. ESRI, Kranzberg.

Flacke, W., 1990b, TIN-EROSION, high-resolution mapping of estimated soil loss, ESRI, Kranzberg.

Grossmann, W.D., 1983a, Systems approaches towards complex systems. Fachbeiträge zur schweizerischen MAB-Information 19, Bundesamt für Umweltschutz, Bern: 25–57.

Grossmann, W.D., 1983b, Systemansatz und Modellhierarchie. In: MAB-Mitteilungen Nr. 16, Ziele, Fragestellungen und Methoden, German national committee of the UNESCO-Programm Man and the Biosphere (MAB), Bonn, 29–33.

Grossmann, W.D., 1989, 'LWMOD Dokumentation zum Landwirtschaftsmodell LWMOD als Grundlage für die Almstudie.' Unveröffentlichter Forschungsbericht, MAB-Projekt 6, Ökosystemforschung Berchtesgaden. (Unpublished.)

Grossmann, W.D., Schaller, J. and Sittard, M., 1983, 'Zeitkarten': eine neue Methodik zum Test von Hypothesen und Gegenmaßnahmen bei Waldschäden. *Allgemeine Forstzeitschrift*, **39**, 834–7.

Haber, W., 1972, Grundzüge einer ökologischen Theorie der Landnutzung. *Innere Kolonisation*, **21**, 294–8.

Haber, W., 1984, Expectations and requirements of ecological research, in MAB-Mitteilungen Nr. 19, Proceedings of the IIIrd International MAB-6-Symposium, April 16–17, 1984, Berchtesgaden, FRG. German national committee of the UNESCO-Programm 'Man and the Biosphere' (MAB), Bonn, 41–54.

Haber, W., 1986a, Mögliche Auswirkungen der geplanten Winterspiele 1992 auf das regionale System Berchtesgaden. MAB-Mitteilungen 22, German national committee of the UNESCO-Programm 'Man and the Biosphere' (MAB), Bonn.

Haber, W., 1986b, Über die menschliche Nutzung von Öystemen – unter besonderer Berücksichtigung von Agrarökosystemen, in *Verhandlungen der Gesellschaft für Ökologie* (Hohenheim 1984), Band XIV, 13–24.

Haber, W., 1990, Using landscape ecology in planning and management, in Zonneveld, I.S. and Forman, R.T.T. (Eds) *Changing landscape: an ecological perspective*, New York: Springer, 217–32.

Haber, W., and Schaller, J., 1988, Ecosystem Research Berchtesgaden – Spatial relations among landscape elements quantified by ecological balance methods. Paper presented at the VIIIth International Symposium on Problems of Landscape Ecological Research Oct. 3–7, 1988, Kamenec Resort Zemplinska-Sirava, CSSR, 1–30.

Kerner, H., 1983, Das Ökologische Bilanzmodell. In: MAB-Mitteilungen Nr. 16, Ziele, Fragestellungen und Methoden, Hrsg.: Bonn: Deutsches Nationalkomitee, 34–50.

Kerner, H., Spandau, L. and Köppel, J.G. (Eds), 1991, *Methoden zur angewandten Ökosystemforschung entwickelt im MAB-Projekt 6 'Ökosystemforschung Berchtesgaden.'* MAB-Mitteilungen 35.1 and 35.2 (2 vols.) Deutsches National-komitee für MAB, Freising-Weihenstephan.

Mesarovic, M., Mackoa, M. and Takahart, T., 1971, *Theory of hierarchical multilevel systems*, New York: Academic Press.

Messerli, B. and Messerli, P., 1979, Wirtschaftliche Entwicklung und ökologische Belastbarkeit im Berggebiet (MAB Schweiz). *Geographica Helvetica* **33**, 203–10.

Messerli, P., 1986, *Modelle und Methoden zur Analyse der Mensch-Umwelt-Beziehungen im Alpinen Lebens- und Erholungsraum. Erkenntnisse und Folgerungen aus dem Schweizerischen MAB-Programm 1979–1985.* Schlussbericht zum schweizerischen MAB-programm Nr. 25, Bundesamt für Umweltschutz, Bern.

Moser, W., 1987, Chronik von MaB-6 Obergurgl In Patzelt G. (Ed.) *MaB-Projekt Obergurgl.* Veröffentlichungen des Österreichischen MaB-Programms 10, Innsbruck: Universitätsverlag Wagner, 7–24.

Rappaport, R.A., 1977, Maladaption in social systems, in Friedman, J., Rowlands, M.U. (Eds) *The evolution of social systems*, London: Duckworth, 49–71.

Schaller, J., 1984, The modelling concept for the MAB-6 project Berchtesgaden. In: MAB-Mitteilungen Nr. 19, *Proceedings of the IIIrd International MAB-6-Symposium*, April 16–17, 1984, Berchtesgaden, FRG. German National Committee of the UNESCO-Programme 'Man and the Biosphere (MAB)', Bonn, 77–92.

Schaller, J., 1985, Anwendung geographischer Informationssysteme an Beispielen landschaftsökologischer Forschung und Lehre, in *Verhandlungen der Gesellschaft für Ökologie* (Bremen, 1983). Band XIII, 443–64.

Schaller, J., 1987, The Geographical Information System (GIS) Arc/INFO, in *Proceedings Euro-Carto VI*, April 13–16, 1987, S. Brno, CSSR. 170–8.

Schaller, J., 1988, Das geographische Informationssystem ARC/INFO. Faulbaum, F., u. Uehlinger, H.-M. (Eds) *Fortschritte der Statisitik-Software* 1. Stuttgart: Gustav Fischer, 503–14.

Schaller, J., 1992, GIS Application in Environmental Planning and Assessment, *Computing, Environmental and Urban Systems*, **16**, 337–53.

Schaller, J. and Spandau, L., 1987, Der Einfluß des Menschen auf Hochgebirgsökosysteme. Integrierte Methoden und Auswertungsansätze zu den Ergebnissen der Ökosystemforschung Berchtesgaden, in *Verhandlungen der Gesellschaft für Ökologie*, Band XV, 15. Jahrestagung Graz, 35–47.

Spandau L. and Tobias, K. (Ed.), 1990, 'Abschlußbericht zur Berglandwirtschaftsstudie – MAB-Projekt 6, Ökosystemforschung Berchtesgaden.' Lehrstuhl für Landschaftsökologie, TU München-Weihenstephan. Unpublished.

4

Land use classification and regional planning in Val Malenco (Italian Alps): a study on the integration of remotely-sensed data and digital terrain models for thematic mapping

Maria Luisa Paracchini and Sten Folving

Introduction

For several decades, research on remote sensing has been directed towards image processing and the development of methods for extracting information, especially on land cover, from remotely-sensed data. The main goals were primarily to produce thematic maps that could be updated quickly. However, the maps obtained from automatic classification of satellite data failed to satisfy their prospective users. They would not readily accept a digital map that could only provide an accuracy of about 75 per cent, even if the new products could provide more classes and cover larger areas than traditional thematic maps.

Computerized geographic information systems (GIS) and remote sensing techniques have been developed in parallel. The growing interest in, and concern for, the environment have created a need for the integration of data to provide information that can be used by planners and politicians in decisions concerning our use of, and impact on, the environment.

In 1985, the Joint Research Centre of the European Communities (EEC) initiated the European Collaborative Programme (ECP) (Folving, 1991; Folving and Megier, 1991) which involved approximately 50 organizations representing both researchers and end users. The main objectives of the ECP were:

- to develop tools and models at the European, regional, and local level for classifying and mapping landscapes by identifying and quantifying their properties;
- to provide spatial geo-information for management decisions and actions for regional development and environmental protection; and

- to analyse and define fields where remote sensing could supply new insight into environmental processes.

The ECP was especially oriented towards the economically less-favoured and marginal areas of Europe, which are often also the most interesting regions from an ecological and environmental point of view. Many of these areas are uplands and mountainous regions which are important hydrological catchment areas.

The study presented in this paper is a part of the ECP, and is being carried out in cooperation between the Environmental Mapping and Modelling Unit at the Joint Research Centre and Instituto di Ingegneria Agraria in Milan. The study is concerned with landscape cover type classification and mapping of Alpine regions for hydrological planning and management. The main aim is to provide a fast and semi-automatic method for assessing the main characteristics of the surface cover in Alpine watersheds.

Digital terrain models are used for geometric correction of the remotely-sensed data: Landsat Thematic Mapper (TM) and SPOT data. Thematic map data are used to aid in classification and for accuracy checking. The results of the classification, together with the thematic map information and terrain models, constitute elements of the GIS that is being built in order to make it possible to carry out modelling studies, primarily on runoff and erosion.

The paper concentrates on two initial parts of the study: image processing (geometrical correction and classification), and map layer production and integration. However, it should be noticed that the approaches are directed by the modelling considerations.

Description of the study area

The Mallero valley is located in the north of Italy (Figure 4.1), within the Alpine chain, and is one of the lateral valleys of the Valtellina, the long east-west valley of the Adda river. The Mallero Valley has an area of approximately 320 km² in a orographically very complex region containing two of the major mountain massifs in Valtellina (Bernina and Disgrazia). The valleys in the region are characterized by very steep slopes, especially in the upper parts which, combined with the heavy rates of rainfall in the area leads to severe slope instability. One of the more famous examples of this was the 1987 landslide (40 million m³) near Sondrio which dammed the upper course of the Adda river and killed 27 persons.

The Mallero river flows into the Valtellina valley at the city of Sondrio. As the Mallero river has caused several floods in Sondrio and carries heavy

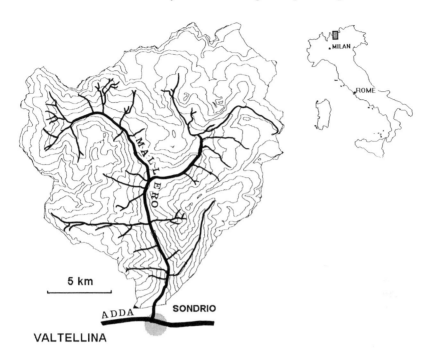

Figure 4.1 The Mallero valley test area.

bed-loads, many regulating dams have been constructed, especially in the lower course but also in the most erosive tributaries.

Geology

Most of the valley is made up of intrusive and metamorphic rocks which are completely impervious (Servizio Geologico). Underground drainage is only possible in the fractured areas where lateral valleys cut the main lithological elements, and in the parts of the valley covered by clastic sediments. The soils are very poor and mainly podsolic.

The morphology of the valley is the result of the combined action of glaciers and rivers. Several glaciers border the watershed: at Mount Disgraza (3678 m), to the west; the Bernina massif (4049 m), to the north; and Mount Scalino (3323 m), to the east. These glaciers regulate runoff over periods of a year or longer.

The Valtellina represents the erosional base for the local catchments; the secondary valley incisions are nearly always transversal and generally very deep with steep slopes. The Mallero valley, in contrast, is much more complex, as its form has been heavily influenced by the local geological structures.

Former valley bottoms are represented by a system of terraces (three in Valtellina), which can also be found in Val Malenco.

Precipitation

The Mallero valley is situated in one of the rainiest areas in Italy (Servizio Idrografico, 1958–1973). The extremely heavy precipitation which often falls in this area frequently causes floods, which is exactly the reason for choosing this area for testing the use of satellite-based land cover mapping for hydrological modelling.

The meteorological conditions sometimes cause very intensive precipitation for long periods; for example, 235 mm fell on Sondrio in 36 hours in September 1964. The number of rainy days is generally higher than 100 per year and the annual total precipitation often exceeds 1000 mm. The rainfall is concentrated mainly in June and November; but the largest amount of precipitation falls, on average, during the autumn months (September to November).

Floods often occur only a few years apart, and sometimes recur during the same year. The Mallero river is one of the Adda river's most important tributaries, and during floods it can carry considerable amounts of solid matter which occasionally cover the alluvial cone. Sondrio has been badly flooded 10 times over the last 100 years. Because the whole Adda basin (2598 km^2) is affected so frequently by catastrophic events – floods and landslides – 33 pluviometric stations have been installed in the area in recent years.

Vegetation

The tree species which grow in the Alpine region are distributed according to altitude and slope orientation, so that it is possible to delimit altitudinal zones or belts of vegetation in which a specific range of vegetation types can be found (Conosci l'Italia, 1958). One can divide a hypothetical mountain slope into basal, montane, and upper zones. The basal zone (0–950 m) is characterized by piedmont vegetation, composed of oaks and chestnuts; the montane zone (950–2250 m) by orophilous vegetation with beech forests, red firs and larches; and the upper zone (above 2250 m) by ipsophyll vegetation mainly consisting of shrubs (rhododendrons and *Pinus montana*) and pastures (Figure 4.2).

In Val Malenco, the vegetation in the zone from 250–900 m consists of mixed broad-leaved forests composed primarily of *Castanea sativa*, *Quercus* spp., *Fagus silvaticus*, *Robinia pseudoacacia*, and *Laburnum* spp. In the zone from 900–1200 m is a mixed broad-leaved and coniferous forest, in which *Fagus silvaticus*, *Acer pseudoplatanus* and *Abies alba* are predominant. Above

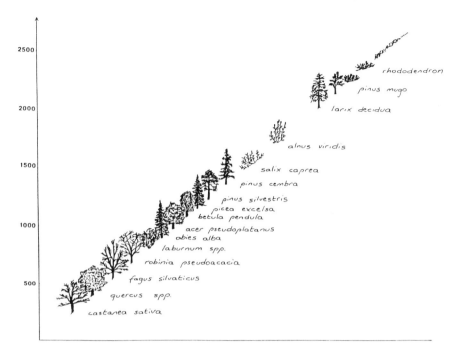

Figure 4.2   Distribution of tree and shrub species in Val Malenco according to altitude.

1200 m, the broad-leaved species tend to decrease in number, leaving space for evergreens (*Picea excelsa, Pinus silvestris, Pinus cembra*) and *Larix decidua*; the only broad-leaved species that can be found up to 2000 m are *Salix caprea* and *Alnus viridis*. Above this height, only scattered *Larix decidua* are found (up to 2200 m) mixed with *Pinus mugo* and rhododendrons. Above 2450 m, the ground is covered by pastures, rocks, and bare soil up to the limit of perennial snow.

As Val Malenco has a north–south orientation, the eastern and western slopes are covered by the same type and density of vegetation. In contrast the vegetation of the east–west lateral valleys is often very different. North-facing slopes have very thick forests, while south-facing primarily have grassland and pastures. These slopes are very rarely covered by forests, and it is not unusual to find isolated trees. The causes are not only differences in micro-climate, but also human impacts: work carried out in the past, primarily deforestation, in order to extend the pastures and cultivated areas. This anthropogenic influence can be seen also in the numerous terraces which can still be found near Chiesa Valmalenco. These areas, taken from the forests in the past, are nowadays neglected and the natural vegetation is reinstating itself.

Problem definition

Model considerations

Operational description and modelling of the runoff and erosion in an Alpine catchment like Val Malenco is best done using a GIS. The model input would be the precipitation, and the output the discharge into the river Adda at the river Mallero's erosional base. Only the composition of the catchment area is considered here, as the aim is to investigate the usefulness of satellite data for this purpose. Many empirical runoff and erosion models exist (Boardman and Favis-Mortlock, 1992; Engman and Gurney, 1991; Morgan and Davson, 1986), with morphology and land cover as the most frequently included characteristics. The morphological parameters needed to characterize the watershed are slope and slope length. The land cover concerns the type of surface and composition of vegetation, if any, and the various ground-covering fractions of the components. Urban and similar areas can be neglected in this context.

For smaller catchments, a very general overflow runoff model can be used (US Army Corps of Engineers, 1976, 1986):

$$R = CA \ (P - D) \tag{4.1}$$

where R is the volume of runoff, A is the area of the catchment considered, P is precipitation, and D is a storage factor. C is a runoff coefficient that expresses the fraction of precipitation that becomes runoff and has to be determined empirically, based on the type and proportions of the surface. This model is best suited for areas covered by solid surfaces, and therefore appears appropriate for bare rock surfaces in mountainous areas.

The land cover classification that is developed here is primarily important for establishing a direct connection to the definition of the curve number (US Soil Conservation Service, 1969) for a given landscape facet. This approach has been taken in several investigations (D'Asaro and Somella, 1988; Jackson and Bodelid, 1983; Ragan and Jackson, 1980; Still and Shih, 1985; Zevenbergen *et al.*, 1988). Using the standard notation for this model, the direct runoff in mm, Q, is given by:

$$Q = \frac{(I - 0 \cdot 2s)^2}{I + 0 \cdot 8s} \tag{4.2}$$

I is storm rainfall in mm and s is the largest potential difference between precipitation and runoff from the beginning of the rainfall s, which is a measure of the soil moisture before the rainfall, depends on the soil conditions and the land cover, and is determined as

$$s = \frac{25\,400}{CN} - 254 \tag{4.3}$$

where the *CN* is a dimensionless, empirical coefficient that describes the partial soil–land cover complexes of the catchment of interest.

Wischmeier and Smith (1978) have provided an estimate of soil loss in tons per hectare per year, *A*:

$$A = RKSG \qquad (4.4)$$

where *R* is a rainfall erosion index stating the annual erosive energy transfer by rain to the surface, *K* is a soil erodibility factor, *S* describes the slope, and *G* is a plant cover describing factor.

These models are very basic compared to the more complicated models used for scientific description of the phenomena. However, they are sufficiently useful as a concept and a framework for development of land cover assessing methods for practical applications. Both the coefficient *CN* and the variables *G* and *S* can be determined, and *K* can be estimated from the digitized soil map.

Geographic information systems

The use of a GIS for storing and manipulating the factors mentioned in the previous paragraphs permits continuous updating and adjustments of the models once the data types have been defined and mapped. While the GIS developed for Val Malenco contains the information necessary for a total assessment of runoff and erosion in the study area, this paper is restricted to the land cover and the digital terrain model (DTM), as this is necessary for the interpretation of the satellite imagery.

Data

Satellite imagery

Because of the heavy cloud cover in mountain areas, it is often very difficult to find useful images for land cover classifications. Furthermore, the period suited for vegetation studies is restricted to the summer months because of the presence of snow until late spring or early summer. Analysis of the availability of images from 1987 from both Landsat-TM and SPOT revealed that the number of images which could be used is very restricted. Of all the recorded scenes, only approximately 10 per cent for TM and 15 per cent for SPOT, generally from the months of August and September, were potentially useful. Consequently, it was not possible to choose the images from the best periods for forestry applications: late May to early June, and, if multi-temporal data sets had been required, August/September. This situation in data availability is believed to be typical for much of the European mountains (Kontoes and Stackenborg, 1990).

The aim of this part of the project is to build a classification methodology, to determine the areas covered by the land cover classes defined as useful for hydrogeological studies. As we are looking for operational applications of the method concurrently with the development, pre-processing work has been minimized. This means that, while atmospheric correction has been omitted, the classification considers shadow effects. Both are very important factors in mountainous regions. It is, however, indispensable to carry out geometric correction of the images chosen for the classification since the relief causes strong parallax effects. When proceeding to the use of multi-temporal images, geometrical correction is not only very important but also an extremely difficult and tedious task. Nevertheless, the aim is not primarily to make highly accurate maps, but to compare the usefulness of different satellite images for operational local hydrological management.

Digital terrain model

In a mountainous landscape like Val Malenco, a digital terrain model (DTM) is necessary for geometrical correction of the imagery; calculation of correct areas, slopes and aspects; and correction for illumination/shadow effects for both optical and microwave data.

The DTM of the study area has been built using the digital contours acquired from the Italian Geographic Military Institute (IGMI) and corresponding to the 50 m contour lines from sheet N18 of the 1:100 000 scale national map series. As the valley is in the north-east corner of the sheet and is not totally included, it was necessary to digitize both the northern and eastern parts of the valley. In the more complex landscape surfaces (in particular the higher summits) where the contour lines were not digitized, it was necessary to complete the digital terrain model by manually digitizing point and additional contours from 1:25 000 scale maps.

The ARC/INFO software package was used to create the DTM of the Mallero watershed (Figure 4.3). The result was transferred from vector to raster format in order to facilitate the necessary further computations with standard and special image-processing software packages.

The most important kind of geometric error is caused by parallax arising from the sensor viewing angle. This parallax causes a shift in the pixel position which is directly proportional to the elevation and distance of the point from the scene nadir. In the study area, the error can reach 300 m for points above 3500 m. Since the scanner records the radiance transversal to the direction of the orbit, the pixels are displaced only in the direction of the scan lines.

The correction for parallax has been performed with specific software developed at the Joint Research Centre (Hill *et al.*, 1989). The procedure consists of three main steps (Figure 4.4):

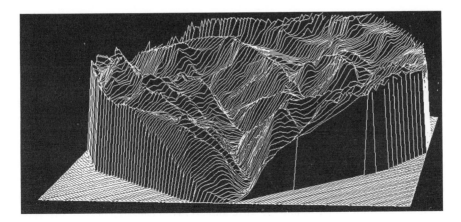

Figure 4.3 Digital terrain model of the Mallero watershed. Integration of manually-digitized elevation data and IGMI contour data processed in ARC/INFO.

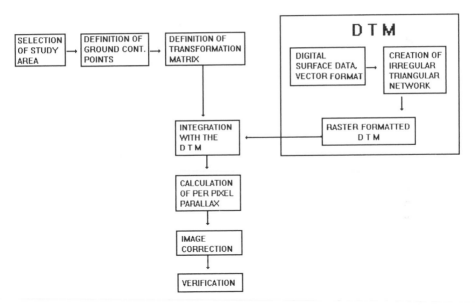

Figure 4.4 Flowchart illustrating the integration of image data with the digital terrain model.

1. a preliminary, rough transformation is performed, using a matrix of coefficients derived from at least ten ground control points;
2. the DTM is integrated for a per-pixel calculation of the distance between the position of the pixels on the image and their correct geographical position;
3. the effective transformation of the image data and resampling of the

radiance data (so-called DN values) is performed, using a 'nearest-neigh-bour' method.

After the correction following the procedure described above, the image still contained a maximum error of approximately two pixels in the areas corresponding to the higher summits. This error has been considered of no importance for the aims of this work, and therefore has been neglected.

Classification method

The image classification is based on the 'Spectral Mixture Modelling' method, originally developed by Adams, Smith, and Johnson (Adams and Smith, 1986; Smith *et al.*, 1990). This method assumes that the pixel reflectance is composed of the linear combination of the reflectances of the elements contained in the terrain surface corresponding to the pixel.

The linear mixture modelling method thus permits the calculation of the percentage of the components whose signals build up the radiance of the pixel. In fact, the vector representing the DN of a pixel in the different spectral bands corresponds to the weighted mean of the radiance (in every band) of all the cover types existing inside that pixel. By applying this method, it is possible to obtain a new vector representing the percentages of the components corresponding to those cover types:

$$x_i = \sum_{i=1}^{n} f_i M_{ij} + e_i \qquad (4.5)$$

where n is the number of bands, x_i is the signal value in band i, f_i is the percentage of pixel area occupied by cover j, the M_{ij} coefficients are called 'end-members' and represent the signal coming from cover j pure pixels, and e_i is a noise term.

The method gives an estimate of f_i using M_{ij} values as input. There are two important constraints:

$$\sum_i f_i = 1 \quad \text{and} \quad 0 < f_i < 1 \qquad (4.6)$$

The first means that the pixel is assumed to be composed only by the cover types under consideration, and the second that the pixel cannot be made up of less than 0 per cent or more than 100 per cent of one determined cover type.

Some errors may occur, affecting the results, due to causes such as:

- non-linear mixing;
- not all the components existing in the image have been considered;
- one or both of the above-mentioned constraints have been ignored;

- there is too much noise in the signal; or
- the end-member values are not well-defined or not adequately separated.

The end-member values must be chosen with extreme care because the results strictly depend on them; the signal should correspond only to one pure component (one cover type). Two of the methods used in the end-member assessment are based on unsupervised and supervised approaches. The first method involves analysis of the image scatterogram in the principal component space; the pure pixels are located on the corners of the distribution of pixel signatures, while the 'inner' values correspond to the different mixing degrees. Once the pure pixels have been identified, it is not difficult to go back to their DN values, used as end-members. The second method is based on ground surveys or aerial photographs, which permit the recognition of the pure pixels on the image.

The spectral mixture modelling technique is particularly suitable for the type of applications foreseen for this study since the intended result is a subdivision of the pixels into classes of vegetation type and density. The classes chosen in the classification procedure are: snow; water; bare soil or rocks; forest; grassland. For the last three classes, intermediate classes were calculated according to variable vegetation density. In practice each pixel was considered to be composed by a variable percentage of two classes x and y, whose distribution has been subdivided according to the following:

4/4 – pixel containing $x > 75\%$ and $y < 25\%$
3/1 – pixel containing $50\% < x < 75\%$ and $25\% < y < 50\%$
1/3 – pixel containing $25\% < x < 50\%$ and $50\% < y < 75\%$

(it was chosen to ignore the presence of pixels made up of more than two elements). Snow, water and glacier surfaces were easily detected and mapped by using maximum likelihood procedures. These areas were masked on the data in the following procedures.

The mixed class bare soil/grassland was treated in a special way. Since unvegetated areas above a certain altitude can be classified to a high accuracy, a distinction could be made between pixels made up of 100 per cent bare rocks and soils and those containing at least 75 per cent. The threshold values used for delineating these surface cover classes were determined from aerial photos and ground surveys. The best result was obtained by treating the bare rock and soil surfaces according to illumination conditions, i.e. to divide these areas into fully illuminated and shaded.

The bare rock and bare soil classes were also used to mask the satellite image into areas fully or partially covered by vegetation. After this second masking of the image, only vegetated areas and their various degrees of mixing with bare rocks and soil are left for further classification.

The classification procedure is a sort of hierarchical analysis, in which one or more classes are extracted and masked at each step (Figure 4.5). Thus,

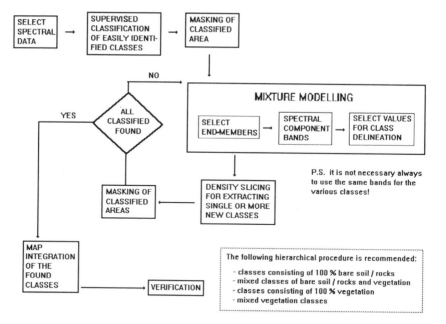

Figure 4.5   Flowchart illustrating the procedure used for spectral mixture modelling and hierarchical classification.

subsequent classification is carried out on the image from which already classified areas have been masked. In practice, at each step, the classes extracted are those which can be identified with greatest accuracy and least effort. The last classes to be extracted were the mixture of forest and grassland, which are more complex to define and classify since shaded or slightly shaded grassland and pastures have reflectance values very similar to those of broad-leaved trees in full sun.

The study has clearly shown that it is worth examining which band combinations that will provide the best results for a given area; it is not necessary to include all bands from a TM scene. The combination of TM bands 2, 4, 5, and 7 could supply all the needed information in Val Malenco. It was also investigated whether principal components analysis would give better results. The first four components extracted from the full data set were used. Both these data combinations were useful and gave almost identical results. The best result was obtained by performing an enhancement of the bands, providing a well-contrasted image using a simple histogram equalization, followed by linear de-mixturing of the enhanced bands, before classification by applying threshold values.

Four end-members were extracted to perform this part of the classification: one for bare soil, one for grassland, and two for forest (broad-leaved and coniferous respectively). This division of forest into two subclasses was necessary in order to better distinguish meadows from broad-leaved trees in

Table 4.1   The areal distribution of the main land cover classes in the Val Malenco area

Class	Area (ha)	Percentage
Forest 3 — grassland 1	3560.4	11.03
Forest 4/4	5981.4	18.52
Grassland 3 — forest 1	402.57	1.25
Grassland 4/4	3370.5	10.44
Forest 3 — bare soil 1	500.94	1.55
Forest 1 — bare soil 3	1342.98	4.16
Grassland 1 — bare soil 3	1279.62	3.96
Grassland 3 — bare soil 1	1071.18	3.32
Bare soil 4/4	876.51	2.71
Rock and urban	10168.56	31.49
Water	148.68	0.46
Shadow	509.4	1.58
Snow	3073.14	9.52

full sun; the results of this distinction were summed to obtain only one forest component. This de-mixing provided a reasonable result, whose validity was checked by airphotos and field reconnaissance.

Results

The results of the process (Plate 2) are summarized in Table 4.1. Further subdivision can of course be performed when, and if, needed. Areas covered by grassland or herbaceous plants constitute 15 per cent of the classes (grass 4/4, grass 3 – bare soil 1, grass 3 – forest 1 are combined). The areas covered by the forest classes, i.e. areas with more than 50 per cent forest cover, comprise 10 042 ha (31 per cent of the total area). The rest of the area has a very sparse vegetation cover, or consists of bare soil, rocks, and glaciers.

The relative distribution according to height is shown on Figure 4.6. Not unexpectedly, areas covered by grasses and other herbaceous plants are found above the tree-line at about 2250 m, but generally the height difference is only approximately 300 m. This is probably due to the fact that former cropland in the valley bottoms is now grassland. There is also a strong tendency for grassland to be found on flatter ground (Figure 4.7). Again this is due to the agricultural practice in the valley. While gentle slopes with some soils development were once turned into pastures, these areas are now being slowly abandoned and are therefore changing into forest. The majority of the forested area is found on slopes between 20 and 40 degrees. The differences in the percentage distribution between forest and grassland on gentle slopes are, as mentioned, due to the land use tradition. This is probably also why mostly forested areas with up to 25 per cent grassland are on somewhat flatter areas

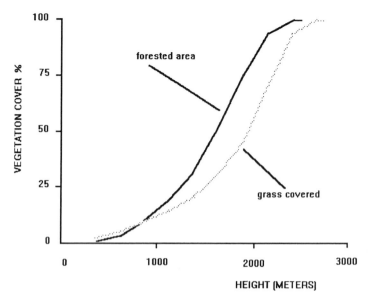

Figure 4.6   Hypsographic curves of forested areas and grassland.

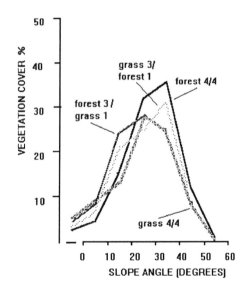

Figure 4.7   Relative distribution of some land cover classes according to slope classes.

than pure forests, whereas grassland with up to 25 per cent trees is mainly found on steeper land.

These types of summaries of information are interesting not just from a geographical point of view, but especially for local landscape planning. Runoff and erosion depend particularly on the combination of surface cover and slope distribution. The surface composition according to height is also interesting, as it is useful for assessing snow cover and snowmelt runoff.

Conclusions

The aim of this pilot study was to improve the application of remotely-sensed optical data for watershed management in the Italian Alps. The results have yet to be integrated fully in the geographic information system, and predictive models have to be established and tested.

The fact that the images were not corrected for shadow effects does not seem to have influenced the results significantly, except that certain very well-defined deeply-shaded areas have been assigned to the 'forest 3 – bare rock 1' class. This error was corrected by image stratification, which removed this class from the surfaces above the vegetation limit.

Shade generally constitutes a severe problem in remote sensing, not least in mountain regions. The signal from shaded surfaces consists generally of a mixture of noise and scattering from the atmosphere and the surrounding pixels; the part of the signal originating in the actual surface is very near to zero. Shadow is not only caused by topography, the various cover types produce their own characteristic shadow. Figure 4.8 shows the total shadow component, extracted by spectral unmixing of the enhanced principal components. The image should have very distinctly revealed the topography by differences in illumination and shadow, but is blurred because of differences in 'internal' shadow of the cover types (especially coniferous and broad-leaved). The pseudoscopic effect is lacking. On level ground, the shadow component pertinent to the specific cover type is an important factor for accurate forest classification. For instance, in the square on Figure 4.8 delineates a homogeneously sloping area, the difference in the shadow component of coniferous and broad-leaved is easily seen; the darker areas at the bottom of the slope are broad-leaved forest. In mountain regions, research is still needed on how to use the shadow information (total shadow minus topographically-caused shadow) for land cover classification.

The integration of a DTM permitted highly accurate semi-automatic surface cover mapping by giving a very precise description of the surface slope structure and the corresponding vegetation cover. Integration of the geological information and soil maps or maps of Quaternary deposits makes it possible to evaluate the runoff even from small land facets.

The land cover part of the GIS database has been completed. The lithologic and pedological information is being integrated – which in itself

Figure 4.8   Shadow abundance derived by applying spectral mixture modelling to enhanced principal components. The internal shadow component of various land cover types often overrides the topographic shadow. For instance, while the area delineated by the square slopes evenly, there is less internal shadow in broad-leaved forest (to the right, on the lower slope) than in coniferous forest (to the left, on the upper slope).

gives rise to many interesting problems, on which the study is presently concentrating.

Acknowledgments

The authors want to express their gratitude to their colleagues in the Environmental Mapping and Modelling Unit at the Joint Research Centre and to Xintie Zhou at SSTC, China, for assistance and support to this study.

References

Adams, J.B. and Smith, M.O., 1986, Spectral mixture modelling: A new analysis of rock and soil types at the Viking Lander 1 site, *Journal of Geophysical Research*, **91**, B8, 8098–112.

Boardman, J. and Favis-Mortlock, D.T., 1992, Soil erosion and sediment loading of water courses, *SEESOIL*, **7**, 5–29.

Conosci l'Italia – Vol. II, 1958, La Flora. Touring Club Italiano. Milano.

D'Asaro, F. and Somella, A., 1988, Uso dei dati telerilevati per la stima indiretta dei volumi di deflusso superficiale. IV Convegno Nazionale, Ingegneria per lo sviluppo dell'agricoltura, Porto Conte-Alghero, 251–7.

Engman, E.T. and Gurney, R.J., 1991 *Remote sensing in hydrology*, London: Chapman and Hall.

Folving, S., 1991, The European Collaborative Programme – a frame for European landscape ecological mapping, *European IALE Seminar on Practical Landscape Ecology*, Roskilde, Vol. 1, 135–43.

Folving, S. and Megier, J., 1991, European Collaboration on development of new environmental thematic map-types from remotely sensed data. Comite Francais de Cartographie. Bulletin No. 127–128, 41–6.

Hill, J., Kohl, H. and Megier, J., 1989, The use of the digital terrain model for land use mapping in the Ardeche area of France, *Proceedings of the 9th EARSeL Symposium*, Espoo, Finland, 236–46.

Jackson, T.J. and Bodelid, T.R., 1983, Runoff curve numbers from Landsat data, Renewable Resources management application of remote sensing, Seattle, May 22–27, 543–73.

Kontoes, C. and Stackenborg, J., 1990, Availability of cloud-free Landsat images for operational projects. The analysis of cloud-cover figures over the countries of the European Community, *International Journal of Remote Sensing*, **11**, 9, 1599–1608.

Maione, V., 1990, The Val Pola (Italy) Rockslide. Management problems of hydraulic emergency, *International Conference on River Flood Hydraulics*, 17–70.

Morgan, R.P.C. and Davson, D.A., 1986, *Soil erosion and conservation*, Harlow: Ingman.

Ragan, R.M. and Jackson, T.J., 1980, Runoff synthesis using Landsat and SCS model. *Journal of the Hydraulics division*, **106**, HY5, 667–78.

Servizio Geologico d'Italia. Carta Geologica d'Italia; foglio 7–18 della carta 1:100 000 dell' IGM, Pizzo Bernina-Sondrio; Foglio 19 della carta 1:100 000 dell' IGM, Sondrio.

Servizio Idrografico. Annali Idrologici, 1958–1973.

Smith, M.O., Ustin, S.L., Adams, J.B. and Gillespie, L.A., 1990, Vegetation in deserts: I, A regional measure of abundance from multispectral images, *Remote Sensing of Environment*, **13**, 1–26.

Still, D.A. and Shih, S.F., 1985, Using Landsat data to classify land use for assessing the basinwide runoff index. *Water Resources Bulletin*, **21**, 6, 931–40.

US Army Corps of Engineers, 1976, Urban storm water runoff, STORM. Hydrologic Engineering Center. Davis, USA.

US Army Corps of Engineers, 1986, Remote sensing technologies and spatial data applications. Hydrological Engineering Center. Davis, USA.

US Soil Conservation Service, 1969, National Engineering Publications, Section 4, Hydrology, Littleton, Colorado, USA.

Wischmeier, W.H. and Smith, D.D., 1978, Predicting rainfall erosion losses. Agriculture Handbook No. 537. Science and Education Administration, US Department of Agriculture. Washington.

Zevenbergen, A.W., Rango, A. Ritchie, J.C., Engman, E.T. and Hawkins, R.H., 1988, Rangeland runoff curve numbers as determined from Landsat MSS data. *International Journal of Remote Sensing*, **9**, 3, 495–502.

5

Towards resolving the problems of regional development in the mountains of the Commonwealth of Independent States using geographic information systems

Alexander Koshkariov, Tatiana M. Krasovskaia and Vladimir S. Tikunov

Introduction

The reader of the articles describing the status of the GIS studies in the former USSR (Koshkarev, Tikunov, and Trofimov, 1989; Koshkarev, 1990), should not bother himself with thoughtful reading because the gap separating these years from today is rather deep, though not so wide as an outsider would expect. 'The collapse of the totalitarian regimes in the now independent states of the former USSR and of its East-European satellites, the processes of democratic revival of their societies, removal from the political arena of the vicious organizations of the communist orientation, more or less successful economic reforms establishing new relationships, demilitarization of the post-communist societies, end of the 'empires of fear', freedom of information and the new information order, transition to civilized forms of relations with the outer world – all these fundamental and inspiring changes . . .' (as a result of the next Russian revolution – 'perestroika' – A.K.) – '. . . have interfered in a profound, and hopefully irreversible way into the patterns of modern life, science, production' (Koshkariov, 1992). If applied to the theme of this chapter, this means the beginning of the end of the totalitarian control of the economic and other aspects of life.

The late 1980s marked the beginning of a period of unique opportunity to revise the established system of resource planning and management, to make new inventories of resources using the modern technologies of remote sensing and geographic information systems (GIS), and to introduce new theories and models of regional development. This chapter describes the mountains of the Commonwealth of Independent States (CIS) and environmental issues of concern for regional development, and presents current

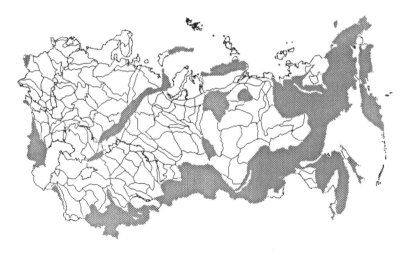

Figure 5.1   Mountain lands of the CIS.

projects, involving GIS at various scales, at the International Mountain Research and Information Laboratory (IMRIL), Moscow, and the Faculty of Geography at Moscow State University.

The mountains of the CIS

The mountains of the CIS cover a vast area (Figure 5.1). These mountains can be divided into five main regions. The first two are in Europe. Furthest west are the Eastern Carpathians of the southwest Ukraine (highest peak: Goverla, 2061 m). To the south and east are the Caucasus, including the highest mountain in Europe, Elbrus (5642 m). The northern side of this chain is in the Russian Federation, including parts of the Krasnodarskiy Region (*Kray*) and the republics of Kabardino-Balkaraya, North Ossetia, Chechenya, Ingushetia, and Daghestan. In addition to Elbrus and Kazbek (5033 m), there are many peaks over 4000 m along the border between the Russian Federation and Georgia and Azerbaijan. To the south of the main Caucasus are the Trans-Caucasian mountains of Georgia (not a member of the CIS), Azerbaijan, and Armenia (highest peak: Aragats, 4090 m). The Urals, with a maximum elevation of 1895 m (Narodnaya), form the divide between the European and Asian (Siberian) parts of the Russian Federation. Far to the north, near the Finnish border, are the low mountains of the Kola peninsula (highest peak: Gora Chasnachorr, 1208 m).

Mountains form most of the southern and eastern borders of the CIS, and cover almost all of the Central Asian states of Tajikistan and Kyrgyzstan and the eastern parts of Uzbekistan and Kazakhstan. The Pamir-Alai of Tajikistan (highest peak: Peak Communism, 7469 m) form the border with

Afghanistan, Pakistan, India, and China. To their north, the Tien Shan (Pobedyi, 7437 m, in Kyrgyzstan) cross the border from Kyrgyzstan and Kazakhstan into China. There are also a number of lower mountain ranges in Kazakhstan, Turkmenia, and Uzbekistan. At the junction of Kazakhstan, China, Mongolia, and the Russian Federation are the Altai mountains, which include the highest peak in Siberia (Belukha, 4506 m).

The Altai provide a transition between the Central Asian mountains and those which cover much of the rest of east and south Siberia, both along the Mongolian and Chinese borders and in the interior, stretching to both the Arctic and the Pacific Oceans. Among these mountain ranges, little known except to scientists and climbers from the former Soviet Union, are the Sayan and Tuva mountains (highest peak: Mongun-Taiga, 3970 m); the mountains around Lake Baikal (Kodara, 2999 m); the north-east Siberian mountains and plateaux, including the Chersky (Pobeda, 3147 m), Verkhoyansky, and Stanovoy ranges; and the mountains of the far east, including those adjacent to the Amur and Okhotsk Seas (Yudomskiy Ridge, 2889 m), the Sikachi-Alin (Tardoki-Yani, 2077 m), Sakhalin (Lopatin, 1609 m), and Kamchatka (Kliuchevskaya Sopka, 4750 m).

Environmental issues

As the mountains of the CIS are at latitudes from arctic to semi-tropical, with climates from maritime to extreme continental, they exhibit a great diversity of physical environments. Similarly, their histories of human use vary greatly. These have been particularly long in the Carpathians, the Caucasus, and Central Asia, all of which have long been characterized by traditional patterns of agriculture in valleys and seasonal grazing in subalpine and alpine zones. However, overgrazing and monoculture agriculture in the mountains of the Caucasus and Central Asia in recent decades have led to major decreases in productivity. Deforestation, leading to soil erosion, desertification, and other environmental changes including catastrophic landslides, has been a major problem in many regions, particularly arid areas, e.g. Uzbekistan and Kirghizia. High rates of population growth, particularly in Islamic republics, also lead to increasing pressures on the environment. Conversely, the emigrations, often forced, of mountain people which took place from the 1930s to the 1970s from the Caucasus, Azerbaijani, and Central Asian mountains have only recently begun to be reversed. In all of these mountain regions, many rivers have been dammed for irrigation and hydro-electricity, and more dams are under construction, mainly in Central Asia, and planned, particularly in the Caucasus. Soil, water, and air pollution is significant in all mining areas, such as the Kola peninsula. Tourism is an increasingly important source of environmental and socio-economic change in the Caucasus and Kirghizia, and also in many other regions.

While the Altai and some of the other southern Siberian mountains have a long history of grazing by the animals of both settled and nomadic tribes, human settlement and use in most of the Siberian mountains have been greatly limited by their severe environments and, particularly, their harsh climates. Only in this century have the mountains of northeast and far east Siberia begun to be developed for mining; forestry has also become a major industry in the far east. While tourism is also growing in the Siberian mountains, numbers of tourists are generally very low, especially because of problems of accessibility. However, tourism is widely seen as a catalyst to economic development, and many projects are underway or planned, especially in the far east (Kamchatka, Sakhalin) and the Altai. In addition, while relatively few dams are currently operational in mountain areas, many are under construction or planned in most Siberian mountain ranges, except Kamchatka where geothermal power is available.

Activities at IMRIL

A preliminary assessment of the introduction of GIS for evaluating the state and management of mountain environments has been undertaken by the IMRIL, which is associated with the Institute of Geography of the Russian Academy of Sciences and the United Nations University. The implementation of GIS technologies to support decision-making for environmental and economic development has been made possible through the availability of digital maps, original statistical data, and field observations. A range of raster and vector GIS have been used to support IMRIL projects: ARC/INFO, TerraSoft, EPPL7, and IDRISI. These GIS have been integrated in long-term regional studies on the sustainable development of the mountain regions of the Caucasus and Central Asia. An early project concerned the evaluation of the environmental consequences of the 1988 Spitak earthquake (Armenia) (Borunov, Koshkariov, and Kandelaki, 1991).

The MACIS project

Mountain Areas of the CIS (MACIS) is a data retrieval, reference, and spatial analysis system utilizing GIS technology. It includes the digitizing of small-scale maps to provide a multi-layered digital base map, and the collection of attribute data related to the features of individual layers. Desktop GIS software is used for data import/export, reference, query, and visualization, and a specialized user interface is used to create interactive links between different data types such as maps, graphics, images, tables, and text. The capabilities of this interface are similar to ArcView or SPANS MAP functionality or hypertext media.

In its first phase (1992–93), the MACIS project includes a 1:5000 000 digital basemap and data layers describing a range of biophysical and

Table 5.1   MACIS digital basemap and data layers

Layer	Source	Number of features
Coastline: seas and inland water bodies (lakes and reservoirs)	USSR base map 1:5000 000	
Rivers and canals	USSR base map 1:5000 000	178 rivers
Administrative boundaries and centres		
– republic (state)	USSR base map 1:5000 000	15 republics
– oblast	USSR base map	171 oblasts
– rayon (for Russia only)	1:2500 000	c. 1400 rayons
Orography		
– linear orographic (morphographic) features	USSR base map 1:4000 000	414 orographic entities
– summits of mountains	various map and text sources	162 summits
Contours (0, 200, 600, 1000, 2000 and 4000 m)	USSR base map 1:8000 000	262 K (.DGT)
Physical-geographical regionalization (including sublayer with province labels)	Thematic map Physical-geographical regionalization of the USSR by N. Gvozdetsky 1:8000 000	19 countries 91 regions 342 provinces
Cartographic grid	Calculation	4° longitude 6° latitude

administrative boundaries and features (Table 5.1). Figure 5.2 shows the data layer of linear orographic features.

Tajikistan-XXI project

Tajikistan is a mountainous state, with altitudes from 300 to 7495 m; nearly half of its area is cold, dry desert above 3000 m. Its deeply dissected relief provides great potential for hydropower, which has been developed since the 1930s. However, the development of hydropower also leads to the large-scale flooding of agricultural land which is in short supply, and the need to move people to newly-developed lowland agricultural areas. While Tajikistan also

Figure 5.2 Orography of the CIS.

has abundant mineral resources, its steep topography and high altitudes mean that agricultural land is limited; there has also been considerable deforestation. The key question for regional development in Tajikistan is whether the environmental and resource capacity will be sufficient to provide living space and meet the demands of a population that is growing rapidly: at a mean annual rate of 3.5 per cent and, in some regions, at 5 per cent (Badenkov, 1992).

An Electronic Atlas-Scenario of Regional Development of Tajikistan in the 21st century (TAJ-XXI) has been planned under the auspices of the UNU Mountain Programme, with the assistance of geographers from Switzerland and the USA. It involves data capture, analysis, and simulation, using various GIS software tools at different spatial levels: macro-level (1:3000 000), meso-level (1:100 000), and micro-level (1:25 000) (Badenkov and Koshkariov, 1990).

The project is illustrated by two examples:

1. a series of small-scale computer maps depicting the basic features of the environment, resources, economy, and population of Tajikistan for each element of its administrative network (the types of transformed landscapes, land use, settlement systems, protected areas, etc.), and
2. the results of meso- and micro-scale data analysis in two test areas: Tajikabad and the Rogoun Hydropower Plant construction area, with digital basemaps, a digital elevation model, drainage network and site locations, and associated attribute data in tabular form.

Macro-level

A set of computer maps of Tajikistan is based on the 46 administrative *rayons* (districts) as the smallest units for data collection and handling (Figure 5.3). A wide range of socio-economic and environmental data is stored for each rayon (Table 5.2). The number of items recorded for each rayon will be enlarged in the future to provide a complete description of Tajikistan, including dynamic data for spatial and temporal analysis. One example from the current database is shown in Figure 5.4, which illustrates the percentage of irrigated land.

Micro-level

There are four test areas for detailed data analysis: Tajikabad, Rogoun, Jagnob, and Sangvor (Figure 5.5). The Tajikabad test area is part of the Vaksh economic region, centred on the Vaksh River. It includes the Garm, Tajikabad, and Jirghital administrative rayons. The principal problems for sustainable development of the mountain geosystems in this region are associated with the heritage of the totalitarian past: i.e. massive forced outmigration to the newly-developed cotton-growing areas of Tajikistan; the consequences of

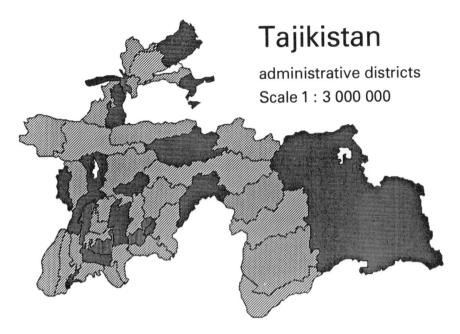

Figure 5.3   Administrative districts (rayons) of Tajikistan.

the Khait Earthquake of 1949, which killed 24 000 people; modern processes of the transformation of animal-breeding practices; depopulation; expansion of abandoned lands; restructuring of land use and slope erosion.

The Rogoun test area (1700 km^2) is in central Tajikistan, between the capital of Dushanbe and the peripheral areas of the Pamirs and Karateghin. It is a zone of transition from middle mountains to high mountains, ranging in altitude from 1150 to 4000 m. The main problems of this area are connected with the construction of the Rogoun Hydropower Plant on the Vaksh river, provoking a fundamental transformation of the natural and socio-economic environment. When completed, the Rogoun groundfill dam will be 335 m high, the highest in the world. From 22 to 40 thousand people may have to be moved from the area flooded by the reservoir.

The Jagnob test area comprises the isolated and narrow Jagnob Valley, over 60 km long in the Pamir-Alai region. It is populated by a small group of Tajiks (300 families), who are direct descendants of the ancient Sogdians forced out in the 7th to 8th centuries AD from the oases of Samarkand and Penjikent to the remote Zeravshan-Turkestan Ridges. They have preserved the Sogdian dialect, and some fragments of ancient traditions, culture, and Zoroastric religion until today. Topics of planned research include: the means and reasons for the conservation of the relic Sogdian culture; the traditional forms of resource management and adaptive mechanisms that have sustained this unique community; and possible ways of integrating this community into the modern world whilst keeping its traditions, culture, and identity alive.

Table 5.2   Attributes (rayon level) in the Tajikistan database

Attribute	Unit
Total area	(1000 km^2)
Total urban population	(1000 pers)
Total rural population	(1000 pers)
Agriculture and forestry	
Total cultivated area	(1000 km^2)
Forested area	(1000 km^2)
Land cultivated outside the rayon by citizens of the rayon	(ha)
Land cultivated within the rayon by citizens from other rayons	(ha)
Irrigated land	(percentage of rayon)
Ploughed and pasture land	(percentage of rayon)
Predominant landscape types	(after N. Gvozdetsky)
Agricultural land per household, Land use class	(percentage of. . . .)
Topography and geomorphology	
Hypsometric curve	(percentage of. . . .)
Mudflow risk	(degree)
Mudflow risk	(season of maximum risk)
Avalanche risk	(degree)
Avalanche risk	(season of maximum activity)
Avalanche risk	(causal factors)

The fieldwork component of the study will involve field mapping and digital data capture using Global Positioning Systems (GPS) and GIS.

The Sangvor test area, in the Obikhingou Valley, faces an expected tourist boom due to the development of the Pamirs National Park. Its planned area would be over 1 million ha. However, there are several problems to solve, including accommodating return migrants from the Vaksh Valley who lived in the region before the 1950s, finding a compromise for hunting Marco Polo goats (licences are currently sold at high cost to western hunters), and finding ways to minimize the impacts of tourism on both ecological and societal systems.

For each test area, a standard set of digital basemap layers is being, or will be, prepared. These will include a digital elevation model (DEM), river network, settlements, and landscape units at scales of 1:100 000 and larger. DEMs for Tajikabad and Rogoun tests have been produced by scanning and vectorizing the 1:100 000 topographic maps in the ARC/INFO format (Figure 5.6). The DEMs for the test areas, primarily created for the ARC/INFO environment, have been translated into the SPANS GIS format with assistance of the Institute of Geography at the University of Berne, Switzerland, for the purpose of producing surface derivatives such as aspect, slope angle,

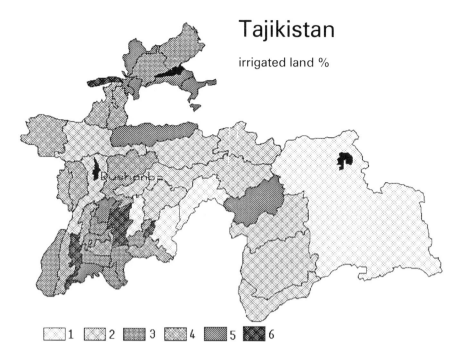

Figure 5.4   Tajikistan: irrigated land (per cent) 1 = 0–3; 2 = 3–10; 3 = 10–20; 4 = 20–30; 5 = 30–40; 6 = >40

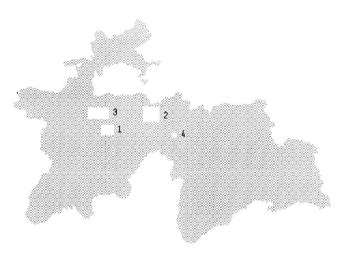

Figure 5.5   Test areas in Tajikistan: 1 = Tajikabad, 2 = Rogoun, 3 = Yagnob, 4 = Sangvor. Field research and a GIS have been prepared for the first two, and are planned for the others.

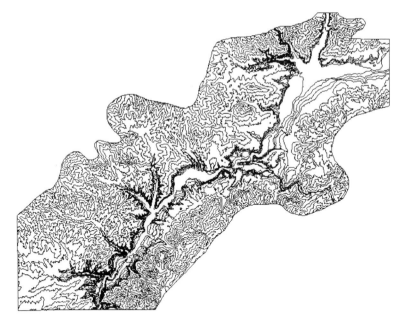

Figure 5.6   Rogoun test area: initial contours as the source for the digital elevation model (DEM), at 40 m and 200 m intervals for the flat and mountainous parts respectively.

three-dimensional (3D) views, and illumination (Figures 5.7 and 5.8). Figure 5.8 shows a 3D view of the Rogoun test area with the surface of the planned hydropower reservoir. This work has revealed several difficulties with the construction and processing (to extract derived morphometric parameters) of DEMs from data sources of varying accuracy without detailed breakline networks.

For a network of settlements (kishlaks), tabular data were collected during the 1992–93 field season. Attributes surveyed for the Rogoun test area included the old and new names; coordinates; altitudes; aspect; topographical position and change of location 1949–90. Administrative status before 1949 and in 1991 were noted together with structure and construction; demography and resources.

Each village was assessed for construction material; shape and growth together with the isolation and clustering of the settlement; seismic-resistant construction and earthquake change in 1949.

Demographic data, official and unofficial, were collected on household numbers before 1949 and in 1990 together with population numbers for 1930; 1958; 1963; 1970; 1979; 1990. Migration, level of depopulation and ethnic composition were also recorded.

The resources of each kishlak were assessed for the economic potential

Stauseeprojekt Rogoun

Geplante Stauhoehe 1240 m

Flaeche ca. 107 qkm

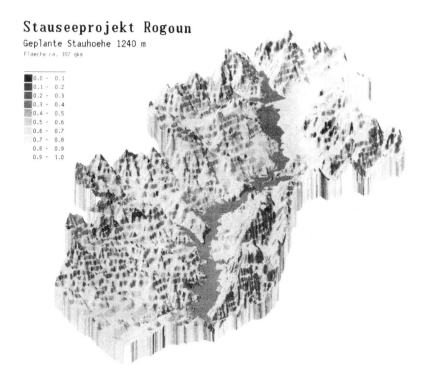

Figure 5.7   Rogoun test area: three-dimensional view, showing the surface (1240 m) of the planned Rogoun reservoir.

of unused land; land ownership pattern; water supply and geodynamic processes close to the village.

A multi-level GIS for environmental studies of the Kola peninsula

The Kola peninsula is the most economically-developed part of northern Russia, with several industrial centres mainly concerned with processing mineral resources. This leads to various environmental problems, particularly pollution and soil erosion. Metallurgical plants in Montchegorsk, Petchenga, and Nickel process copper-nickel ores rich in sulphur, resulting in severe pollution by heavy metals and sulphur dioxide. Kirovsk is one of the largest centres of the apatite-nepheline industry in Russia, with three processing factories and six mines (mainly open-pit) in and around the city. The main resulting problems are air pollution and land reclamation. Kandalaksha is well-known for its aluminium industry, which causes heavy pollution. Thus, the environment of the Kola peninsula experiences a substantial anthropogenic load, and several impact zones can be recognized.

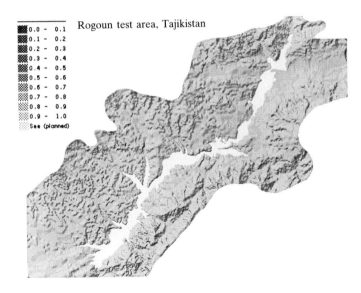

Figure 5.8   Rogoun test area: Illumination map. The 1240 m surface of the planned Rogoun reservoir is also shown.

Environmental studies of the peninsula are of primary importance not only for Russia, but also for other northern countries, because of the need to develop fundamental approaches for the rational exploitation and conservation of northern resources. It is important to standardize and organize the considerable volume of environmental data from the region; GIS are very useful for these purposes.

A problem-oriented environmental GIS has been developed for the Kola peninsula. It has three levels, each including different types of environmental data, from the very detailed to the general. This GIS is a further development of the Ecoarctic GIS (Krasovskaia and Tikunov, 1991a), whose practical application for resource evaluation and the development of a monitoring system has been presented by Krasovskaia and Tikunov (1991b). The three levels of the GIS are the entire peninsula, the Khibiny mountains, and the city of Apatity. The first two levels are presented in this paper, and have potential for both regional planning and nature conservation.

The Kola information system

The first level of the GIS covers the whole of the Kola peninsula. It includes environmental data based on the categorization of physical-geographical regions (natural regions) at a scale of 1:1000 000, developed by Kurbatov (1980) for the eastern and central parts of the peninsula and extended to the western part during the development of the Kola Information System.

The choice of natural regions as the basic spatial unit lies both in the need to incorporate environmental data into the GIS (data at larger scales are often inadequate) and also in the appropriateness of this level of regionalization for compiling data for large areas. Each of the 18 natural regions is defined in terms of its geomorphological and climatic characteristics and has typical combinations of vegetation and soil types. The principal variables which define the regions, relief and climate, are also particularly important factors for the dispersion and accumulation of pollutants.

The environment of each natural region is defined in terms of 17 variables, divided into two main groups. The first group characterizes the environmental potential for the dispersion and accumulation of pollutants, and includes the atmospheric pollution potential (wind velocity, occurrence of temperature inversions, calms, etc.), hydrothermal potential, relief, intensity of biochemical cycling, and pollutant concentrations. The second group comprises different environmental-economic and social variables, such as the index of industrial environmental hazard, the size of industrial centre, and the period of industrial development. The information on these variables that is compiled in the database is derived from field research, mathematical modelling of pollutant transport, and literature analysis.

After the compilation of the database, the rank of each region was determined, using a synthetic evaluation of the 17 variables. The method used was the estimation algorithm developed by Tikunov (1985). This consists of the normalization of the initial data using the following equation:

$$x_{ij} = \frac{x_{ij} - x_j}{\max/\min x_j - x_j} , \begin{array}{l} i = 1, 2, \ldots, n \\ j = 1, 2, \ldots, n \end{array}$$

where n = number of regions (18); m = number of variables (17); x = the worst value for each variable for monitoring purposes (i.e. maximum values for variables such as height above sea level and atmospheric pollution potential, minimum values for intensity of biochemical cycling and height of the convective boundary layer); and $\max/\min x$ = extreme values (most different from x).

A comparison of the 17 variables across the 18 regions in terms of the worst values of x permitted the regions to be ranked according to the intensity of the variables. The chosen method used Euclidean distances as a measure of proximity to a typical region with the worst values of x for all characteristics. Euclidean distance was chosen because other measures did not provide understandable results. The method involved principal components analysis of the information array in order to derive orthogonal groups of variables. The data set was compressed by excluding the last component, which accounted for a negligible proportion of the variance, from further calculations. The resulting Euclidean distances were then used as integral estimation parameters.

The estimation algorithm also permits the identification of groups of similar regions, defined in terms of rank values of the Euclidean series. This multivariate procedure enables the identification of homogeneous groups defined by different groups of variables. The degree of separation was evaluated with canonical correlation coefficients (Griffith and Tikunov, 1990) and heterogeneity coefficients (Tikunov, 1985a), in order to identify a statistically-optimal classification. The final classification comprised four groups (Figure 5.9).

The four groups of regions are defined according to the degree of existing and possible environmental conflicts. The Montchegorsk and Kandalaksha conflict areas can be clearly seen in Figure 5.9. However, this also identifies several parts of eastern Kola where industrial development would lead to a high risk of environmental conflict. As well as the first-order maps, second-order derivative maps have been compiled. One of these is shown as Figure 5.10, which shows the anthropogenic load on natural geosystems. In general, this GIS level is valuable for the planning of economic development by regional authorities.

The Khibiny mountains GIS

The Khibiny mountains are an important centre for mining and processing apatite-nepheline ores. This region is one of the most severely polluted in northern Russia and, being close to the metallurgical plant at Montchegorsk, also experiences pollution from this major source. Anthropogenic change in the Khibiny mountains derives from both mechanical disturbance (quarries, rock waste) and air and water pollution. This is dominated by Sr and P from local sources, but also includes Cu, Ni, and SO_2 from Montchegorsk.

The first difficulty to solve in the development of the problem-oriented GIS for the Khibiny mountains was the definition of hierarchies for the collection and storage of data. The solution was a modular scheme, which can be easily modified for different purposes. The system permits data storage, manipulation, search and visualization, database expansion and protection, modelling, and various outputs. Spatial data may be structured in various ways.

River basins are a common spatial framework for the collection of environmental data. However, the river-basin approach was not appropriate for this work because of the need to compile specific environmental and geochemical data. Consequently, spatial data were structured according to hierarchical landscape units, using the 1:50 000 map of the landscape units of the Khibiny mountains published by Zuchkova (1972). This approach is also highly suitable for developing revegetation plans for anthropogenically-disturbed and polluted areas. The classification of landscape units is based on both vegetation type and geomorphological characteristics, such as plateau tops, slopes, river banks and valleys, and lake bottoms. This topographic differentiation is very important with regard to the accumulation of pollutants.

Figure 5.9 Environmental conflicts associated with industrial development of the Kola peninsula.

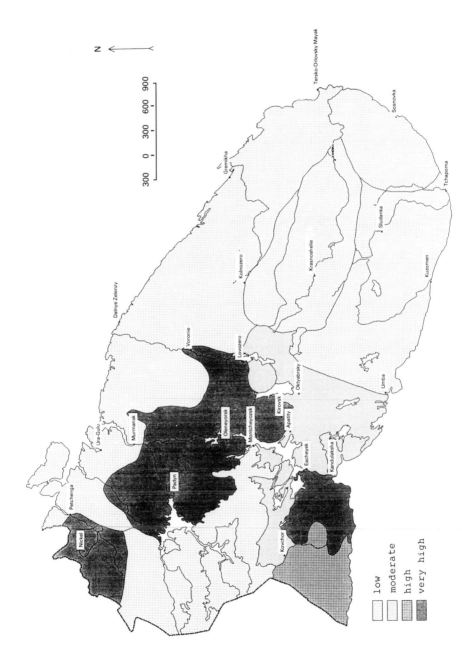

Figure 5.10 Anthropogenic load on natural ecosystems of the Kola peninsula.

Unpublished research by A.V. Yevseev and A.S. Myagkova has shown that concentrations of pollutants in these landscape units depend on altitude, aspect, and vegetation type.

A range of environmental data was recorded for each landscape unit. Altitude is particularly important in relation to aspect; landscape boundaries on south- and southwest-facing slopes are higher. This altitudinal division of landscape also permits the differentiation of pollution-sensitive northern taiga from more stable tundra. This finding comes from the unpublished research of Yevseev and Myagkova, who found an inverse, but non-linear, relationship between height and degree of air pollution. Water quality was defined in terms of hydrochemistry: total soluble matter and concentrations of HCO_3, Na, Cu, Ni, and F in surface water. These data were collected during field research by L.E. Zidyreva and M.B. Zaslavskaia. The occurrence of Ni and Cu reflects long-range atmospheric transport from Montchegorsk.

The remaining descriptors were qualitative, and reflect possible pollution loads. It is planned to replace these classifications with more objective and/ or quantitative measures. Two classes of exposure of a landscape unit to local air pollution sources (apatite-nepheline processing plants, quarries) were defined: no pollution and some pollution. This approach was used because information about local atmospheric circulation processes is inadequate. The information for this categorization derived from field research and expert evaluation. A two-class scale was also used to record the presence or absence of local sources of pollution, including settlements, waste deposits, and drilling sites. The final variable was the type of land use. Six classes were defined:

1. unused areas;
2. recreational zones, parks, and botanical gardens;
3. roads;
4. railways for the transport of apatite-nepheline ore and concentrate (connected with dust pollution);
5. settlements; and
6. industrial sites, quarries, garages, etc.

The methods described in the previous section were used to derive synthetic maps for the Khibiny mountains. Figure 5.11 is the map of anthropogenic loading on landscapes. This shows four classes of loading, of which the highest are associated with the city of Kirovsk and waste reservoirs. The GIS was also used to compile maps of pollutant concentrations, and also of existing and potential environmental conflicts. The latter were based on the analysis of the data described above for specific landscape units. These maps are essential for the current planning of economic development, which must consider not only the mineral but also the recreational resources of the Khibiny mountains. Thus, the work described in this section is valuable for nature conservation planning, the optimization of land use and mining development, environmental impact assessment, and the formulation of environmental

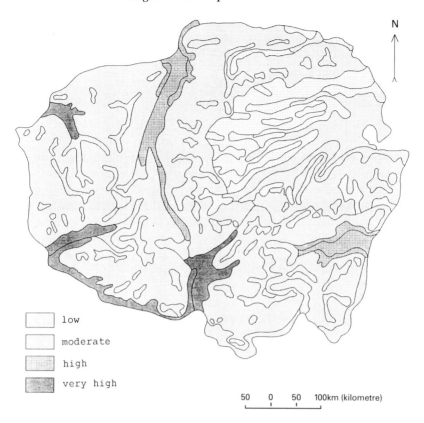

N
↑

low
moderate
high
very high

50   0   50   100km (kilometre)

Figure 5.11   Anthropogenic loads in the Khibiny Mountains, Kola peninsula.

monitoring programmes. It is important to stress that the compilation of environmental data at this scale is very laborious and requires considerable interdisciplinary field research, which is underway at present in the region.

Future developments

The first version of a hierarchical GIS for environmental purposes in the Kola peninsula may be used for regional planning, nature conservation, and environmental monitoring. It permits the inclusion of many different characteristics which may be used to define 'real' environmental quality. A GIS permits the complex combination and subsequent analysis of such characteristics, and finally the prediction of potential environmental conflicts. The flexible approach to data storage used in the database is valuable for different types of environmental study. The landscape unit approach seems to be promising for nature conservation and environmental monitoring. The existing databases will be expanded with new data both for the regions already studied and for new regions.

Conclusions

While the implementation of GIS for regional development in the mountains of the CIS is at an early stage, considerable progress has been made in spite of limited hardware, software, personnel, and training; experience is growing rapidly. GIS is a valuable technology for combining the results of new field research with the experience, often recorded on detailed thematic maps, of many scientists from throughout the CIS, in order to assess the status quo and develop scenarios for the future development of the mountains of the CIS. One should, however, note that many of these regions are zones of conflict as adjustments are made to a new political and economic system, so that GIS is unlikely to be implemented so readily in these regions. Yet, in general, awareness of the potential of GIS is growing in both research and administrative institutions, and this potential will be increasingly realized in the future.

Acknowledgments

A. Koshkariov expresses his gratitude to his colleagues in IMRIL, who provided him with data that were used for this article, to Yu. Badenkov who was the organizer and inspirator, to P. Messerli, B. Messerli, H. Gerhandinger and to all staff of the Geographical Institute of the University of Berne who spared no efforts and expertise for computerization of the study of Tajikistan.

References

Badenkov, Y., 1992, Mountains of the former Soviet Union: value, diversity, uncertainty, in Stone, P.B. (Ed.) *The state of the world's mountains*, London: Zed, 257–97.

Badenkov, Y. and Koshkariov, A.V., 1990, Development of new techniques for resource evaluation and development: an electronic atlas for Tajikistan, in *Making it work, GIS '90 symposium proceedings*, Vancouver: Forestry Canada, 67.

Badenkov, Y.P., Koshariov, A., Klassen, V., Gounia, A., Merzliakova, I., Piesnoukhin, V. *et al.*, 1993, Tajikistan: scenarios of development in a diverse natural environment, society and economy, in Badenkov Y.P. *et al.* (Eds) *Transformation of Mountain Environments* (in press).

Borunov, A.K., Koshkariov, A.V. and Kandelaki, V.V., 1991, Geoecological consequences of the 1988 Spitak earthquake (Armenia), *Mountain Research and Development*, **11**, 19–35.

Griffith, D.A. and Tikunov, V.S., 1990, Comparative analysis of modelling algorithms for the content of typological maps, *Geodezia i cartografia*, **8**, 39–43 (in Russian).

Koshkarev, A.V., 1990, Cartography, geographic information and ways they interact, *Mapping Sciences and Remote Sensing*, **27**, 185–98.

Koshkariov, A.V., 1992, Geographical Information Systems in the CIS: a critical view on the critical state, in Cadoux-Hudson, J. and Heywood, D.I. (Eds) *Geographic Information 1992/3*, London: Taylor and Francis, 35–40.

Koshkarev, A.V., Tikunov, V.S. and Trofimov, A.M., 1989, The current state and the main trends in the development of geographical information systems in the USSR, *International Journal of Geographical Information Systems*, **3**, 257–72.

Krasovskaia, T.M. and Tikunov, V.S., 1991a, Evaluation of the ecological situation in sub-polar regions with the help of geoinformation technology, *Environmental change and GIS*, **2**, 243–50.

Krasovskaia, T.M. and Tikunov, V.S., 1991b, Geographical Information Systems for environmental impact assessment, in *Proceedings, Second European Conference on Geographical Information Systems*, **1**, EGIS Foundation, 574–80.

Kurbatov, N.L., 1980, Physical-geographical regionalization of the Kola Peninsula, *Vestnik Moskovskogouniversiteta seria geographia*, **N2**, 50–2, (in Russian).

Tikunov, V.S., 1985, Classification methods for geographic complexes in making evaluative maps, *Vestnik Moskovskogouniversiteta seria geographia*, **N4**, 28–36, (in Russian).

Tikunov, V.S., 1985a, *Modelling in Social-Economic Cartography*, Moscow: Moscow University Press, 280 pp.

Zuchkova, V.K., 1972, Landscape map of the Khibiny Mountains, in *Landshaftnoe kartografirovanie i fiziko-geografícheskoe rajonirovanie gornich stran*, Moscow: Moscow University Press, 43–52 (in Russian).

6

Geographic information systems as a tool for regional planning in mountain regions: case studies from Canada, Brazil, Japan, and the USA

Kurt Culbertson, Bonny Hershberger, Suzanne Jackson, Steve Mullen and Heidi Olson

Introduction

In North America, geographic information system (GIS) technology has largely been utilized by academic institutions and governmental entities charged with the planning and management of large land areas and governmental juris-dictions. However, the private sector is increasingly using GIS as a tool to address regional planning questions. The case studies presented in this chapter illustrate opportunities, pitfalls, and solutions that Design Workshop has found in using GIS in mountain environments in Canada, Brazil, Japan, and the USA (Figure 6.1). Design Workshop is a mountain-based land-planning and landscape-architecture firm, specializing in the planning, design, and man-agement of projects involving tourism, mining, protected areas, and urban development.

A broad range of computer technologies, including GIS, computer-aided drafting and design (CADD), image processing, and three-dimensional simulation, has been used in these case studies. They exhibit an evolution in the application of GIS from visual resource assessment and comprehensive environmental analysis, to economic development issues and, finally, to the application of the sustainable development model to planning.

Bow-Canmore Visual Impact Assessment, Alberta, Canada

Each year, 3·3 million visitors visit Banff National Park in the Rocky Moun-tains of Alberta, Canada. Most of them travel through the 33 km Bow-Canmore valley, east of the Park (Figure 6.2). The valley's scenic beauty, characterized by the juxtaposition of man-made developments and natural

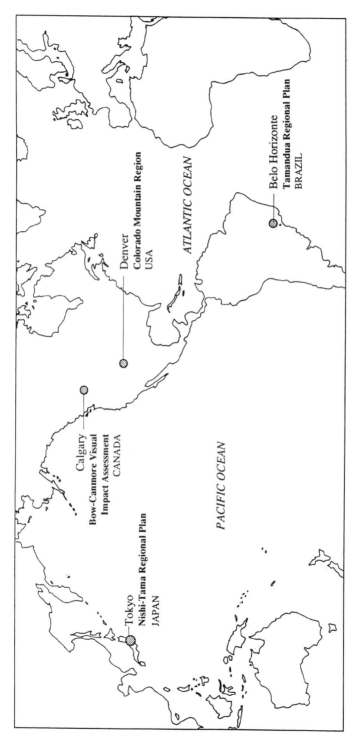

Figure 6.1 Locations of case studies.

Figure 6.2   Location of Bow-Canmore study area.

scenery, is recognized as the key component of the valley's visual character. The natural scenery has been moulded by glaciers and, more recently, the Bow River, which flows 1800 m below peaks such as the 3000 m high Three Sisters.

In recent years, there have been increasing numbers of proposals for resort, mining, industrial, and commercial development in the area. These threaten to change the visual quality on which the tourism industry, a strong economic factor in the region, is based. In response, Alberta Tourism co-ordinated the Bow Canmore Tourism Development Framework in order to examine key issues and provide specific information and recommendations. One objective was to define a method of maintaining the visual quality of the environment in relation to future development. To meet this goal, the Bow Canmore Visual Impact Assessment was developed (Alberta Tourism, 1991). It is based on a methodology that goes beyond classical visual analysis (e.g. Smardon *et al.*, 1986) by combining the use of computer technology, for the inventory and analysis of information and the simulation of alternative futures, with the involvement of a public group for input on visual quality and acceptable levels of visual modification to the landscape. The assessment resulted in a map that illustrated Visual Quality Objectives for the Bow-Canmore Valley and an associated set of Design Guidelines.

A GIS was considered critical for developing the four-stage methodology for this regional-scale project and applying information based on pho-

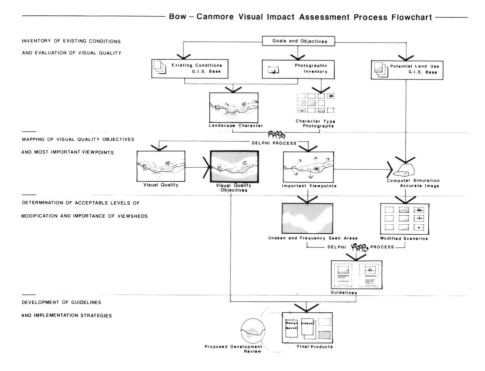

Figure 6.3   Bow-Canmore Visual Impact Assessment process flowchart.

tographic views to two-dimensional maps. As shown in Figure 6.3, the four tasks were:

- to inventory existing conditions and evaluate visual quality;
- to map visual quality objectives (VQOs) and establish most important viewpoints;
- to determine acceptable levels of modification and importance of viewsheds; and
- to develop design guidelines and implementation strategies.

The first task, the inventory of existing conditions, began with the establishment of a computer database using the GRASS (Geographic Resources Analysis Support System) GIS, run on a Silicon Graphics IRIS workstation. The goal was to inventory all information that contributed to the visual quality of the area, including existing and potential land use. The base information came from a variety of sources. Elevation, hydrology, and transportation maps were transformed from ARC/INFO files to GRASS, while maps from Alberta Transportation, Alberta Tourism, and Alberta Forestry, Lands and Wildlife (Steber 1990; O'Leary, 1988), were tablet-digitized to create thematic maps including potential development; zoning; ecology; and areas of scenic significance. In all, 10 base data maps were developed. Base data at 1:20 000 or 1:15 000 (25 m raster grid size) provided a sufficient level of detail for the 33 km-long study area and, in some instances, very detailed classes of information were combined. The study area boundaries were not based on political jurisdictions, but were defined by visual boundaries as experienced from the valley floor. Most of these were ridgelines, which were identified using the GIS.

During an on-site photographic inventory, slides and photographs, taken primarily from major transportation corridors and high-use areas, were classified according to their landscape characteristics. These ranged from natural areas, such as rivers, lakes, and rocky mountain peaks, to man-made developments, such as major roads or hamlets. In all, 13 landscape categories were identified using the photographic views. This information was then applied to a two-dimensional map of landscape character by using the GIS to reclassify basemaps such as elevation; water; and existing land use.

Next, recognizing that visual quality defined by experts is virtually identical to visual quality as defined by the general public (Zube, 1974), a group of 21 lay persons familiar with the area was called upon to rate the slides of the area on a scale from very ugly to very beautiful. The Delphi method (Brown, 1968) was used to facilitate consensus. This led to the second task of mapping VQOs and establishing most important viewpoints. This used the GIS to apply the visual quality ratings and objectives established by the public group. The visual quality map describes the visual significance given to the landscape and comprises six classes from very beautiful to very ugly. It was created by reclassifying the landscape character map by incorporating the public visual quality ratings. For example, coniferous forests were classified beautiful, while rocky, vegetated slopes were classified moderate to ugly. This map, in turn, was reclassified to create the VQO map which describes the degree of acceptable alteration of the characteristic visual landscape (Figure 6.4).

The VQO categories are:

- full protection: only ecological change is permitted; this should apply to rivers, lakes, and high mountain peaks;
- partial protection: human interaction should be unobtrusive to residents or visitors and should not detract from the natural scenic quality; this should apply to coniferous forests, mixed vegetation, meadows, reservoirs, and minor impacts to water and vegetated areas;
- modification and improvement: human developments may dominate the characteristic landscape, but should use natural form, line, colour, and texture, and appear similar to natural features when viewed in the foreground; this should apply to major roadways, rocky vegetated slopes, towns, hamlets, and commercial development along highways; and
- improvement: the scenic value and quality of existing conditions should be enhanced; this should apply to areas affected by mining.

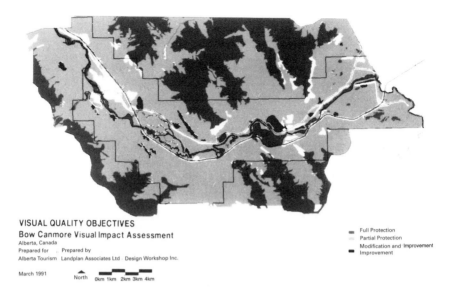

VISUAL QUALITY OBJECTIVES
Bow Canmore Visual Impact Assessment
Alberta, Canada
Prepared for   Prepared by
Alberta Tourism   Landplan Associates Ltd   Design Workshop Inc.

March 1991   North   0km 1km 2km 3km 4km

▬ Full Protection
   Partial Protection
   Modification and Improvement
▬ Improvement

Figure 6.4   Bow-Canmore Visual Quality Objectives map.

The public group also identified important viewpoints in the study area in a cognitive mapping exercise. Eight viewpoints, including the scenic Grassi Lakes and the Main Street of the town of Canmore, were identified in addition to the major transportation corridors. There were two main reasons for identifying these viewpoints: first, to locate areas of the valley that are either hidden or frequently seen from important viewpoints and, second, to provide locations from which to simulate the effects of change. In this instance, computer analysis capabilities proved invaluable in combining landform, slope, and elevation information to generate viewshed maps which showed the areas that could be seen from specific viewpoints. Several of these viewshed maps could then be combined to identify which areas were most frequently seen and therefore more visually sensitive, and which were hidden from view and therefore more appropriate for visually-unobtrusive development. Although the computer program was so time-consuming that it had to be run overnight, this technology showed the mountainous landscape in a way that was otherwise almost impossible to achieve.

The third task was to determine acceptable levels of modification. Another type of computer technology provided a tool which facilitated communication and public input. A three-dimensional modelling system was used to create alternative scenarios of future development. These scenarios were superimposed over scanned photographs and electronically painted to create realistic images (Figure 6.5). The resulting images were then evaluated by the public group on a scale from worst to best in relation to existing conditions. The images were also used as a basis for group discussion on alternative

Computer simulated views

Upper Left: Existing Conditions
Lower Left: Middle to Background Lowrise Residential Development

Upper Right: Foreground Roadside Strip Development
Lower Right: Combination of Foreground Roadside Strip Development and Middle to Background Low Rise Residential Development

Figure 6.5   Computer-simulated views of alternative scenarios of development.

futures for the region. This led to the development of design guidelines and recommendations for implementation.

This final task included comparing existing land use policies with the VQOs. This would have been extremely laborious using traditional overlay techniques. In fact, a GIS database presents such a vast array of opportunities for analysis that a landscape architect must be very selective in choosing information that is critical to the project. In this study, ecological land suitability information for development was compared with the VQOs. First, the ecological land classification map was reclassified using ecological characteristics, to identify areas in terms of suitability for development. The resulting 'land suitability' map was then compared with VQOs using a matrix which permitted the identification of areas of supportive, neutral, or conflicting visual impact. For example, areas of low development potential were considered to have supportive visual impact when they were also recommended for protection. However, high development potential would lead to conflicting visual impact in most cases, other than where the VQO of improvement was recommended. The result shows a high degree of compatibility between ecological land suitability and VQOs. However, when comparing development potential based on zoning policy with the VQOs, there are many areas with apparent conflicts. More than 4800 ha were identified where decision-makers should review existing land use policy in order to reduce potential visual impact conflicts.

The strength of this study is in the representation of VQOs that are based on collective values. This was achieved by using computer technology to integrate public input with two- and three-dimensional representational systems. Another special factor was in responding to the client's needs by creating computer-generated information that could be used on an ongoing basis. This has been achieved; the Province of Alberta is presently using the data, originally generated on GRASS and TARGA systems, on SPANS and MIPS. In addition, the client required explicit documentation so that the visual assessment methodology could be applied to similar mountain valleys and serve as a prototype to guide visually-sensitive land use in the Bow-Canmore area. Finally, a slide presentation was prepared which explains the intent and conclusion of the study to the public.

In producing the final products, several things became apparent.

- The use of integrated computer systems strongly influenced the methodology for this multi-faceted study. At the same time, the process had to be clearly and simply communicated to the client;
- It was too easy to generate too many maps, once the GIS database was established. It was critical to identify which maps were essential and to select the appropriate format for the maps from the wide range of output possibilities;
- It is important to be aware of the limits of the technology. For example, as the project and data are at a regional or sub-regional scale, the results

should only be applied at that scale. If a question arises that is site-specific, these products should only be used for discussion or to raise a red flag.

- The public expects computer-generated maps to have a graphic quality and detail of information that is as good or better than those produced by hand. The quality of the final output depends on the information put into the computer as well as hardware, software, and output capabilities, all of which are continually improving. The latter are often determined by economic feasibility.

Among the lessons and opportunities learned were:

- to allow the functioning of the computer to influence the development of an appropriate methodology for the project;
- to use the capabilities offered by computer technology, but keep the process clear and understandable for the client and the public;
- to understand the limitations of current technology in order to create products that satisfy the needs of the client and are economically feasible;
- to prioritize digital products relative to the goal of the study, in order to determine appropriate types of final output;
- to assist the client in following the process so that he can maximize use of the study's final products and database; and
- to continue to enhance and develop land analysis capabilities, for example, comparison of land use policies or viewshed projections.

Tamandua regional plan, Brazil

The lessons learned from the application of GIS to questions of visual resource management were then applied to a planning project with broader environmental concerns. Design Workshop, in conjunction with its Brazilian subsidiary, was commissioned to study redevelopment alternatives for Aguas Claras, a 900 ha open-pit iron ore mine adjacent to Belo Horizonte, Brazil (Figure 6.6), which is owned and operated by Mineracoes Brasieriras Reunidas, SA.

An early part of the project was to study the transition from mining to village, including plans to direct the configuration and location of the remaining mine waste deposits. During the planning process, a variety of computer applications were utilized. First, a GIS was created for the entire property to guide long-term management. Later, simulation techniques were used to visualize the various stages of the mining plan, and to illustrate the three-dimensional form of the real estate development. The project resulted in a new village that included clustered mixed-use development in a pedestrian-oriented and transportation-efficient scheme as a model for a new Brazilian satellite community. The use of traditional architecture and forms of historic

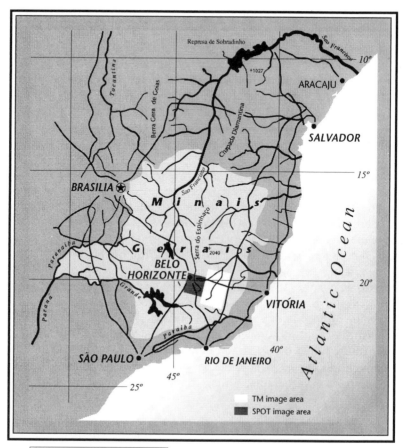

Locator Map

The Belo Horizonte Region
Brazil, South America

Figure 6.6   Location of the Belo Horizonte region, Brazil.

local villages creates a place that fits the region and culture. About half of the property was set aside as a forest preserve.

Following the completion of Aguas Claras, work began on an expanded master plan for the 35 100 ha Tamandua region, also near Belo Horizonte, which is owned by a consortium of Brazilian mining companies. It lies mostly within the Quadrilatero Ferrifers, a mountainous region known for its deposits of iron ore. The goal of the project was to determine which areas were suitable for mining or urban development, and which should be set aside as environmental preserves. The master planning process involved the design of an inventory database and analysis management tool that could be used by the landowners to maximize the profitability of their land in a manner that would be responsive to the public welfare.

At the outset of the design process, it was determined that, because of the scale of the project, a GIS database would be the most efficient and effective way to handle the volume of data needed to produce a comprehensive and usable product. It was also essential to be able to easily update base information and to change decision criteria to produce alternative development models.

The available base information existed in a variety of forms and scales. Topographic maps ranged from hand-drawn maps of mining sites to 1:50 000 government maps. In order to reduce the time spent on data entry, the topography was scanned from the 1:50 000 maps. Political boundaries, roads, and hydrology were then hand-digitized from these maps, and geology, from 1:25 000 US Geological Survey maps. The use of data from different sources made registration a challenge. Not only were the maps at different scales, but many were photocopies. Print stretch created further problems. Through a proprietary addition to GRASS, a best fit type of stretching of the vector information was performed in order to register each layer. Once all the information was digitized, an elevation model was generated from the topographic file. Watershed basins, aspect and slope information (Figure 6.7), and a viewshed analysis were derived from the elevation model.

Although vegetation maps were available, most were very old, hand-drawn, of indeterminate scale, and lacked reference points with which to connect the data to a coordinate system. Consequently, land cover types had to be derived from satellite imagery. In order to maximize the accuracy of this information, a SPOT image (20 m resolution) was preferred to a Landsat-TM image (30 m resolution). The best fit process was repeated in order to match the satellite imagery with the vector data. The first stage of creating a land cover map from the satellite imagery was to isolate the different wavelengths in each of the three SPOT bands within a small area of the image with known vegetation types. This permitted the classification of the entire image by grouping the wavelengths which corresponded with known vegetation types. However, some types of vegetation, though known to exist in the area, were difficult to identify using the satellite imagery. In order to increase the accuracy of the land cover map, the information was plotted and

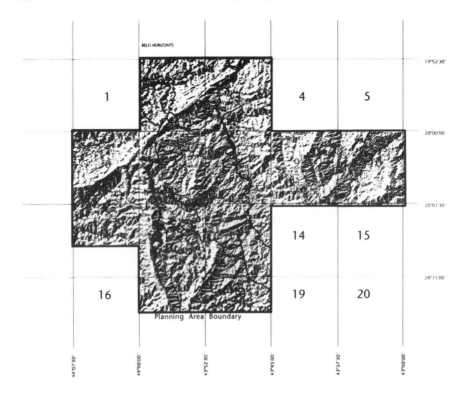

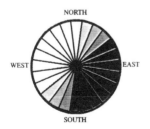

LEGEND

Figure 6.7   Aspect map of the Tamandua region, Brazil.

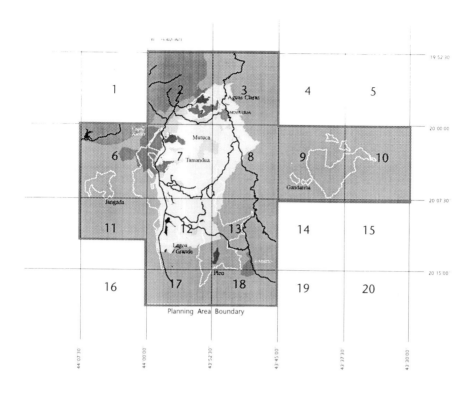

LEGEND
Cities/Towns
Nova Lima
Preserves/ Parks
Lakes / Reservoirs
Existing MBR Mines
Condominiums

Figure 6.8 Land use map of the Tamandua region, Brazil.

field checked. Information for areas which could not be clearly classified by utilizing the SPOT data was supplemented by digitizing the vegetation information derived from field survey.

.Inventory maps were generated for each layer of information in the database. These maps included hydrology, geology, land cover (Plate 3), visual corridors, watershed basins, slope, political boundaries, and major roads. This information allowed the systematic analysis and identification of the capability of the land for real estate and urban development.

The first step in the analysis process was the combination of information on land cover, geology, slope, and proximity to provide an environmental summary map with three categories: highly sensitive, moderately sensitive, and not sensitive. Though development should minimize visual impact on the landscape, the visual analysis was not taken into consideration as a restricting factor at this phase in the planning process. This information will be used in a later, more detailed phase of the planning process.

After generating the environmental summary map, a masterplan/land use map (Figure 6.8) was created. Areas identified as environmentally sensitive were set aside as preserves or parks. Areas of moderate and low sensitivity were then combined with other criteria to identify which areas were most suitable for commercial and industrial centres and high- and low-density residential development, and which had the greatest mining potential. The resulting map is now being used as a guide in identifying areas of potential urbanization and preservation. In summary, the GIS database formed the basis for development of a master plan based on a comprehensive environmental impact statement for the area.

Nishi-Tama regional plan, Japan

Having applied GIS to a broad range of environmental issues in the Tamandua project, our next goal was to incorporate economic considerations into the decision-making process for land use. Design Workshop was selected to participate in a symposium in Tokyo to demonstrate a planning process and recommend concepts for the development of the Nishi-Tama region, 30 km west of the Tokyo metropolitan area (Figure 6.9). The assumption was that, as resort planners from the West, we could provide a seasoned perspective for planning the future of this newly-emerging tourism region. As Tokyo continues to grow as the focus of government and business activity at both national and global scales, the Nishi-Tama region is assuming a unique and important role in the future of the Tokyo metropolitan area. The planning project aims to help to shape a healthy social, environmental, and economic future for the five communities of Nishi-Tama, and consequently, to begin to suggest the region's long-term relationship with Tokyo.

With Japan's growing support for an environmental ethic and responsible stewardship of the land as major components of a rewarding lifestyle, and Nishi-Tama's proximity to Tokyo, the region has the possibility of becoming a Central Garden Park, promoting preservation/conservation-based tourism development. The proposed goals were to promote the conservation of the pristine views and natural resources, encourage sustainable economic development, and preserve the history, culture, and communities of the region.

The process of environmental inventory and analysis was used to support and develop the conceptual plan. This was an integral part of our communication at the symposium, at which we stressed that the first step of

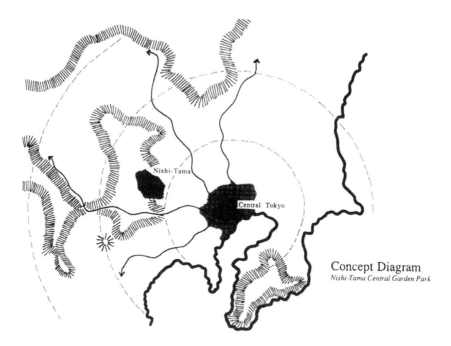

Concept Diagram
Nishi-Tama Central Garden Park

Figure 6.9   Location of the Nishi-Tama region, Japan.

environmentally responsible planning is to understand the land capabilities. Computer technology was used to provide a clear and simple explanation of how to analyze natural systems to determine environmental constraints, and how to identify the most- and least-developable areas. Three-dimensional terrain models were particularly valuable in communicating to a public unaccustomed to reading maps.

In an illustrative initial study, MacGIS software was used to inventory and analyze existing information for two of the five communities in the region. Slope, elevation, and aspect information was digitized from the extensive topographic maps available. Information on vegetation types was collected and tablet-digitized. Information on soils and floodplain was not available, and was therefore fabricated based on gross estimates of the exposure, valley bottoms, and slopes. With this information, it was important to emphasize the use of the inventory and analysis maps only as tools to explain the process, and not as a definitive determination of land capability.

We felt that environmentally responsible planning needed to create opportunities for economic development while emphasizing environmental and visual concerns. Although this resulted in different, valid criteria conflicting with each other, they could be evaluated within the same GIS framework by developing economic, environmental, and visual capability maps. These were weighted and combined to prepare a development capability summary

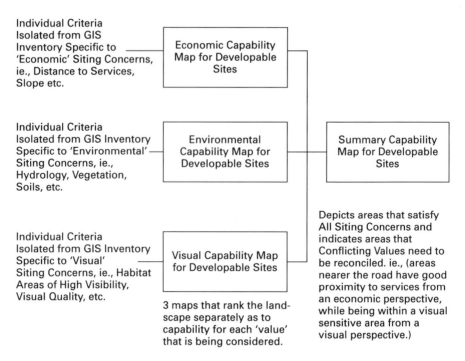

Individual Criteria Isolated from GIS Inventory Specific to 'Economic' Siting Concerns, ie., Distance to Services, Slope etc. ——— Economic Capability Map for Developable Sites

Individual Criteria Isolated from GIS Inventory Specific to 'Environmental' Siting Concerns, ie., Hydrology, Vegetation, Soils, etc. ——— Environmental Capability Map for Developable Sites ——— Summary Capability Map for Developable Sites

Individual Criteria Isolated from GIS Inventory Specific to 'Visual' Siting Concerns, ie., Habitat Areas of High Visibility, Visual Quality, etc. ——— Visual Capability Map for Developable Sites

3 maps that rank the landscape separately as to capability for each 'value' that is being considered.

Depicts areas that satisfy All Siting Concerns and indicates areas that Conflicting Values need to be reconciled. ie., (areas nearer the road have good proximity to services from an economic perspective, while being within a visual sensitive area from a visual perspective.)

Figure 6.10   Development capability methodology for the Nishi-Tama region, Japan.

suggesting one possible outcome of the analysis (see Figure 6.10). This summary was then used to prepare the conceptual development plans for two communities, indicating appropriate development zones and uses based on land capability (Figure 6.11).

Areas depicted as most capable by the combination of the three capability maps were obvious locations for development. However, as these areas were limited, pressure to develop other areas, for instance, along roadways and within natural vegetation, will be inevitable. The real objective was to demonstrate that multi-resource management implicitly requires the consideration of multi-values, so that conflicts between objectives need to be resolved in a socially equitable fashion. Rationales need to be developed to prioritize the choice of which values to emphasize in different locations. For example, environmental concerns should take precedence when future development is proposed within undisturbed areas of natural vegetation, while economic concerns may predominate in urban or developed areas. In the design phase, the team will reconcile development conflicts. Two examples under consideration are the restoration of natural systems along riparian corridors in urban areas, to heal past damage and form a connected natural ecosystem; and visual screening and buffers in visually sensitive areas, to protect scenic views.

This project illustrated the following points:

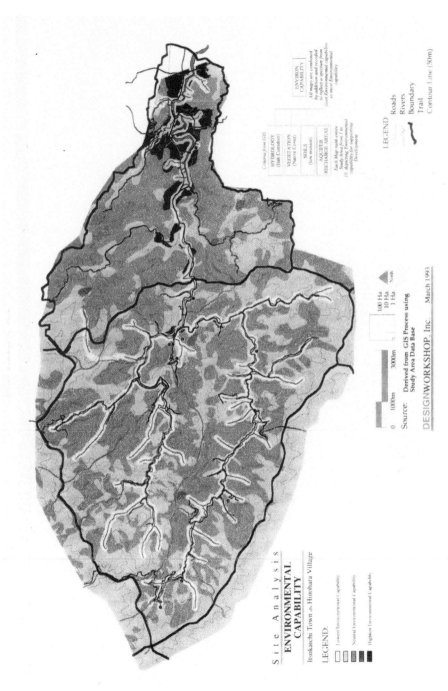

Figure 6.11 Environmental capability map for the Nishi-Tama region, Japan.

- There was great value in describing the planning process using GIS, for evaluating and comparing not only environmental, but also visual and economic conditions. This method of understanding the relationship among the diverse factors that comprise the human landscape can help in making decisions for a sustainable future.
- The MacGIS process was a quick and easy way to prepare maps that clearly describe the process of inventory and analysis of environmental, economic, and visual issues.
- An audience unfamiliar with GIS planning processes understands the simple inventory, analysis, and combination of maps required to prepare a developable land summary. More complex details, such as weighting of different factors, need not be included in an explanation of the process.

The transformation of the Colorado mountain region, USA

The lessons learned in addressing environmental and economic factors through GIS are now being expanded to include cultural considerations. As an ongoing research effort, Design Workshop is developing a comprehensive GIS database for the entire mountain region of Colorado. The goal is to provide a planning tool for directing the development of this region as its economy changes from one predominantly based upon mining and agriculture to one that is largely dependent upon tourism and second-home development. It is hoped that the mapping and combined analysis of environmental, economic, and social factors will contribute to achieving this transformation in a sustainable development context which finds a balance between these concerns.

Utilizing US Geological Survey topographic data, a series of basemaps of slope, aspect, and elevation have been created. These layers have been combined with floodplain and hydrologic data from the US Army Corps of Engineers, agricultural soils data from the Colorado Department of Agriculture, and property ownership data from county and federal agencies to gain an understanding of the potential for development in Colorado's various mountain valleys.

SPOT satellite imagery will be used to monitor land use change over time. It is anticipated that analysis of data from different years will permit observation of changes of land use from cultivated farmland to housing and urban development, as well as changes in timber harvesting. This type of land use monitoring is particularly important because few land management entities in Colorado conform to watershed or other natural boundaries. There are over a dozen state and federal agencies with jurisdiction in the region, including the US National Park Service, the US Forest Service, the Bureau of Land Management, the Colorado Land Board, the Colorado Water Conservation Board, the Colorado Department of Transportation, and the

Colorado Division of Wildlife. Each utilizes different systems of management units and data collection. Consequently, compiling baseline information and monitoring the total landscape change occurring in a given valley is particularly difficult. These problems are further complicated when county and municipal jurisdictions are also taken into consideration. It is not uncommon for a valley within the Colorado Rockies to be divided among seven counties, or for one county to encompass portions of two or three valleys.

In addition to mapping environmental factors, we are attempting to establish economic criteria which can be used in both local and regional decision-making. Real estate locational factors that relate to ski resorts and other tourism facilities are being determined. Farm-to-market considerations for agricultural production and timber and mineral extraction criteria are also being evaluated.

As a third point in the sustainable development triangle, cultural and social factors are also crucial. Using geo-referenced US Department of Census data, both the growth of population and changing demographic characteristics can be monitored for each valley. This information will be important in anticipating and directing the rapid gentrification of these valleys due to the massive influx of second-home owners. A further use of the GIS database would be to combine the developable land summary and demographic data with a water production and use model for each watershed to provide a predictive model for calculating the carrying capacity of each valley.

Conclusions

The four examples from Canada, Brazil, Japan, and the United States offer evidence of the range of problems to which GIS are being applied. A number of conclusions and challenges can be identified.

- Despite the advances in computer and satellite technology, planners must still contend with difficulties in assembling appropriate data for decision-making. The most sophisticated satellite imagery must often be combined with hand-drafted maps and anecdotal information.
- Many applications of GIS technology have unfortunately involved little more than mapping exercises. There is a great challenge for planners to develop the analytical skills and planning methodologies to match the capabilities of the hardware and software.
- In weighting data layers, there appears to be great potential in combining the training of the scientific expert, with the practical experience of the lay person, as in the case of the Bow-Canmore Visual Analysis.
- While there is great potential to utilize GIS technology in the analysis of environmental conditions, its greater value may lie in planning for sustainable development. The potential of combining economic and social information with environmental data offers an exceptional tool for long-term decision making.

Acknowledgments

The projects presented in this paper were undertaken for the following clients: Bow-Canmore Visual Impact Assessment for Alberta Tourism (in collaboration with Landplan Associates, Ltd.); Tamandua regional plan for Mineracoes Brasieriras Reunidas, SA (in collaboration with Design Workshop/Santana); Nishi-Tama regional plan, for Kites Corporation.

References

Alberta Forestry, Lands and Wildlife, 1990, 'Bow Corridor Local Integrated Resource Draft Plan', Edmonton: Alberta Forestry, Lands and Wildlife.

Alberta Tourism, 1991, 'Bow-Canmore Visual Impact Assessment Report', consultants' report, Calgary and Aspen: Landplan Associates Ltd. and Design Workshop.

Brown, B.S., 1968, *The Delphi Process: A Methodology Used for the Elicitation of Opinions of Experts*, Santa Monica: The Rand Corporation.

O'Leary, Dennis, J., 1988, *Integrated Resource Inventory and Evaluations of the Bow Corridor Integrated Resource Planning Area.* Land Information Services Division, Alberta Forestry, Lands and Wildlife, Edmonton, AB.

Smardon, R.C., Palmer, J.F. and Felleman, J.P., 1986, *Foundation for Visual Project Analysis*, New York: John Wiley & Sons.

Steber, J., 1990, Planners Update: *Bow Corridor Local Integrated Resource Plan.* Resource Planning Branch, Alberta Forestry, Lands and Wildlife, Edmonton, AB.

Zube, E.H., 1974, Cross disciplinary and intermode agreement on the description and evaluation of landscape resources, *Environment and Behavior*, **6**, 69–89.

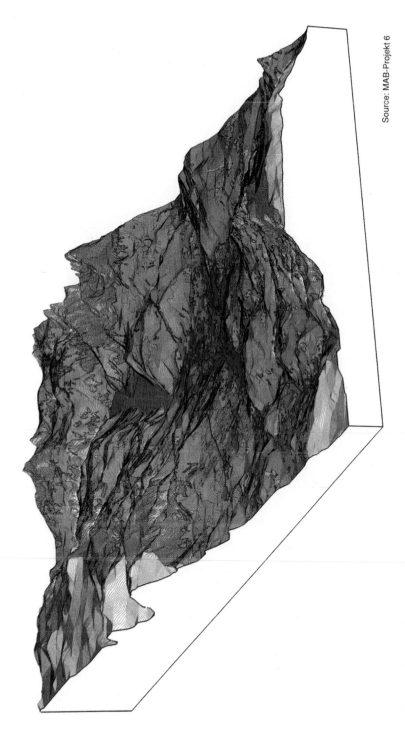

Source: MAB-Projekt 6

Plate 1   Three-dimensional view of the Berchtesgaden area, Germany. (Chapter 3, page 49)

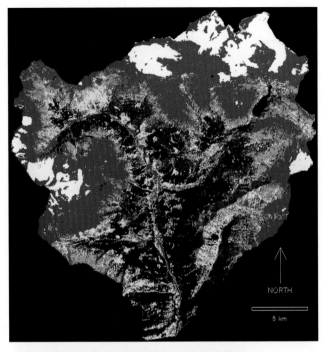

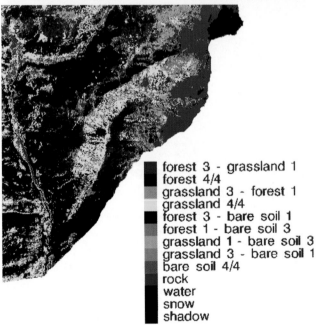

forest 3 - grassland 1
forest 4/4
grassland 3 - forest 1
grassland 4/4
forest 3 - bare soil 1
forest 1 - bare soil 3
grassland 1 - bare soil 3
grassland 3 - bare soil 1
bare soil 4/4
rock
water
snow
shadow

Plate 2   Final landcover map of the Val Malenco catchment area, derived by the method described in the text from a Landsat Thematic Mapper image from 18 August 1990. (Chapter 4, page 71)

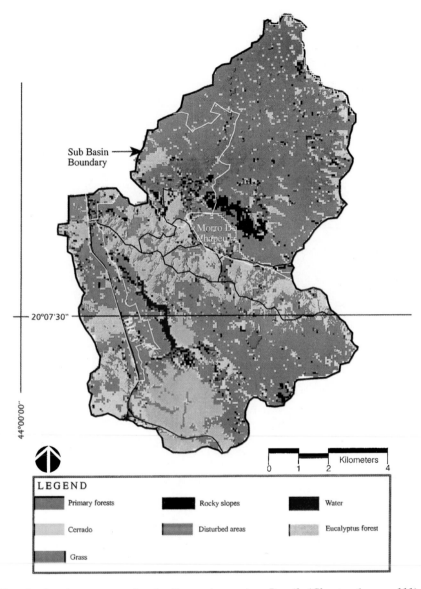

Plate 3   Land cover map for the Tamandua region, Brazil. (Chapter 6, page 111)

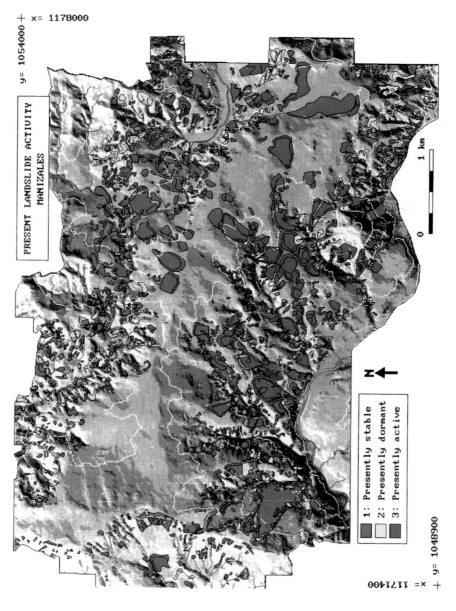

PRESENT LANDSLIDE ACTIVITY
MANIZALES

x= 1178000

y= 1054000

1: Presently stable
2: Presently dormant
3: Presently active

N

0   1 km

x= 1171400

y= 1048900

Plate 4   Landslide activity map for the large-scale study area, Rio Chinchina catchment, Colombia. (Chapter 8, page 148)

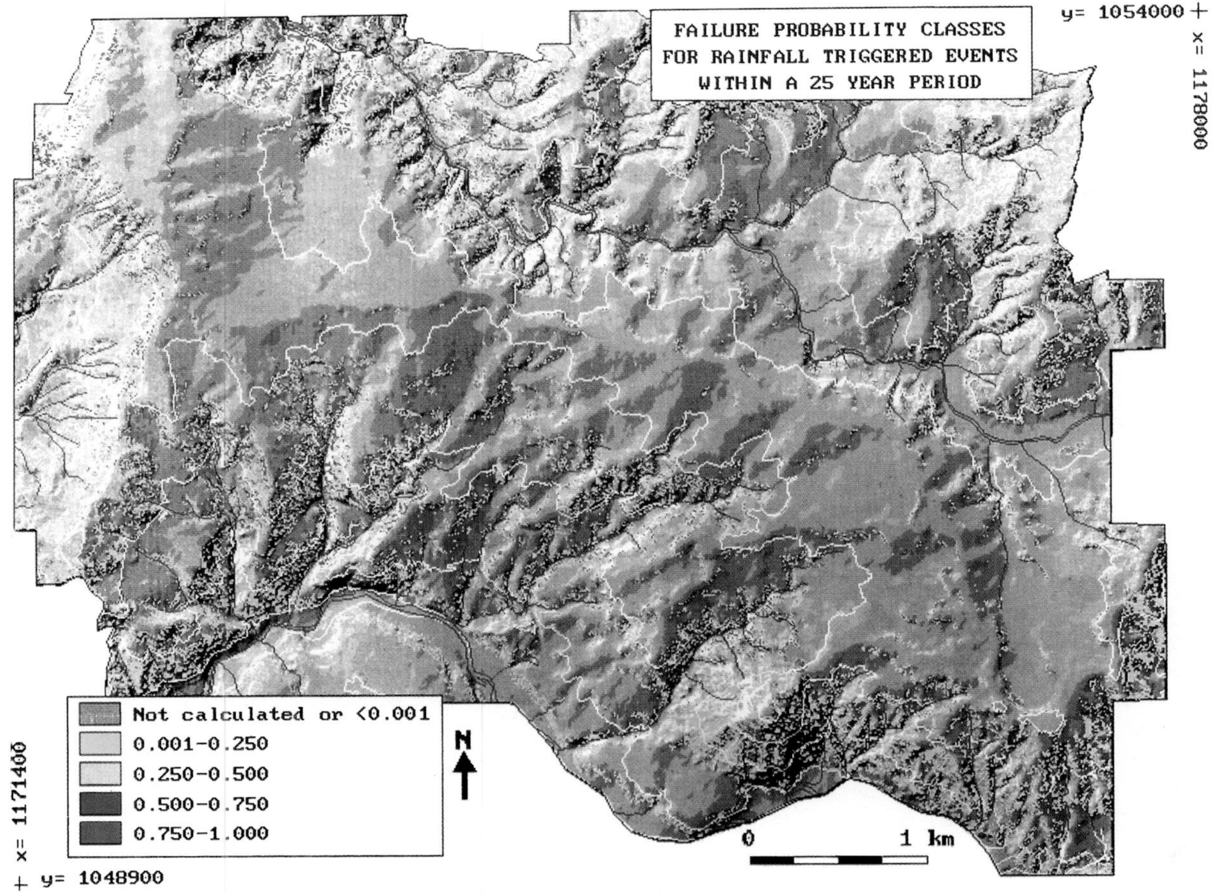

Plate 5   Failure probability for a design period of 25 years related to rainfall-triggered translational landslides, largescale study area, Rio Chinchina catchment, Colombia. (Chapter 8, page 157)

Plate 6   A Landsat Thematic Mapper image composite of Glacier National Park, Montana, USA. (Chapter 10, page 192)

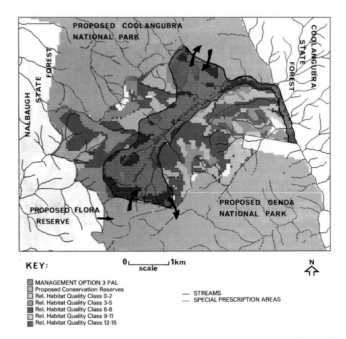

Plate 11   Nalbaugh Special Prescription Area (SPA) Population Assisting Link (PAL). Option 3 PAL between proposed Genoa and Coolangubra National Parks overlaid on a simplified map of relative habitat quality classes. The relative habitat quality map is based on an original containing 15 relative habitat quality classes. (Chapter 11, page 229)

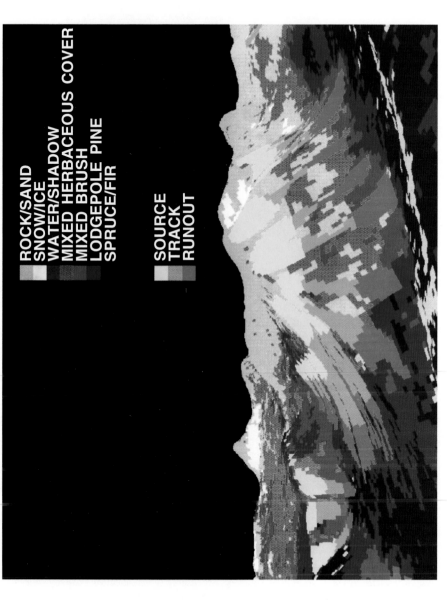

Plate 7 Three-dimensional drape of terrain, landcover types, and snow-avalanche paths achieved, respectively, through a 1:24 000 base-scale DEM, Landsat TM digital data, and digitization of photointerpreted path locations within a representative portion of the Park, Montana, USA. (Chapter 10, page 198)

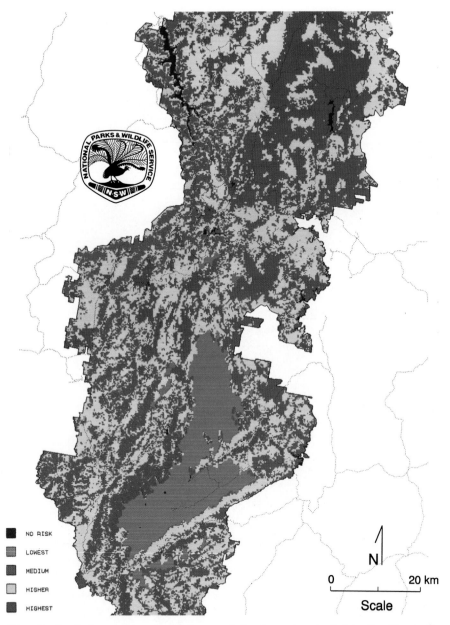

■	NO RISK
▨	LOWEST
▨	MEDIUM
▨	HIGHER
■	HIGHEST

N

0   20 km

Scale

Plate 8   Preliminary assessment of general fire hazard potential in the Kosciusko National Park, based on a spatial analysis of vegetation type, elevation, aspect, and slope. Further factors can be added to the model including rainfall, time since last fire, and fuel weights. (Chapter 11, page 223)

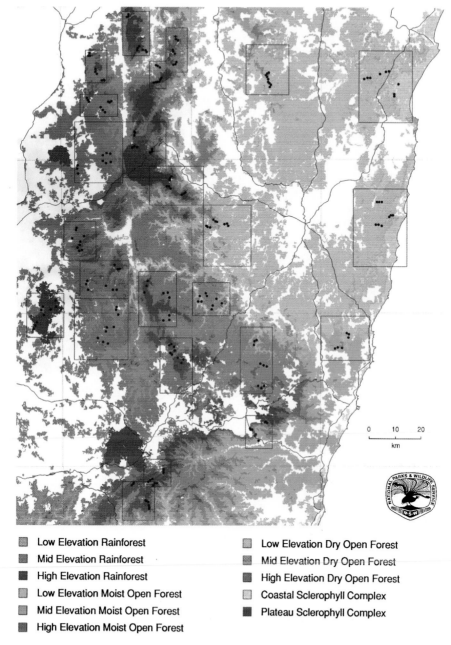

▦ Low Elevation Rainforest	▦ Low Elevation Dry Open Forest
▦ Mid Elevation Rainforest	▦ Mid Elevation Dry Open Forest
▦ High Elevation Rainforest	▦ High Elevation Dry Open Forest
▦ Low Elevation Moist Open Forest	▦ Coastal Sclerophyll Complex
▦ Mid Elevation Moist Open Forest	▦ Plateau Sclerophyll Complex
▦ High Elevation Moist Open Forest	

Plate 9   Location of fauna survey sites in a section of the northeast forests study area. The sites were stratified against combinations of vegetation (derived from LANDSAT TM imagery), elevation, and geology using a GIS database developed for this project. Major roads are also shown. (Chapter 11, page 226)

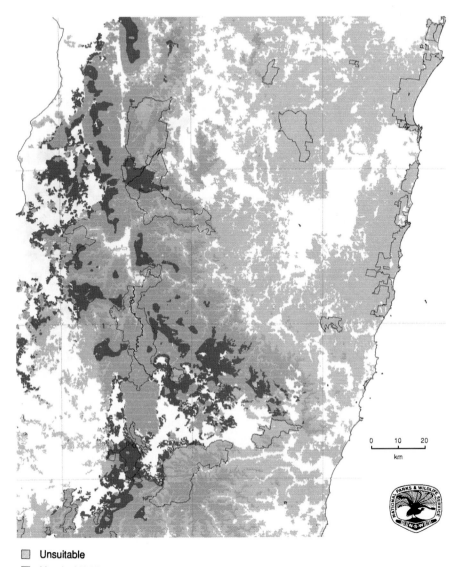

- ☐ Unsuitable
- ▨ Marginal Habitat
- ▩ Core Habitat

Plate 10   Predicted habitat for the Hastings River Mouse, Pseudomys oralis, *in a section of the Northeast forests study area. The prediction is based on a decision tree model derived by analysing correlations between survey data and terrain, climate and vegetation variables held in a GIS database developed for this project. Boundaries of existing conservation reserves are also shown. (Chapter 11, page 227)*

Simulated Map of the Potential Natural Forest Vegetation of Switzerland

Brzeziecki, Kienast and Wildi 1993

Map legend

1	12
2	13
3	14
4	15
5	16
6	17
7	18
8	19
9	20
10	21
11	22

50 km

Plate 12 Simulated map of the potential natural forest vegetation of Switzerland. The resolution of the map is 1 km. The units are: 1. Luzulo-Fagion; 2. Eu-Fagion (brown mull); 3. Eu-Fagion (rendzina, submontane); 4. Eu-Fagion (rendzina, low-montane); 5. Cephalanthero-Fagion; 6. Abieti-Fagion; 7. Aceri-Fagion; 8. Tilio-Acerion; 9. Alno-Fraxinion; 10. Carpinion; 11. Quercion pubescenti-petraeae; 12. Quercion robori-petraeae; 13. Piceo-Abietion; 14. Molinio-Pinion; 15. Vaccinio-Piceion; 16. Larici-Pinetum; 17. Erico-Pinion; 18. Dicrano-Pinion; 19. Sphagno-Pinetum; 20. urban areas; 21: lakes and wetlands; 22. alpine environments (after Brzeziecki et al., submitted). (Chapter 14, page 267)

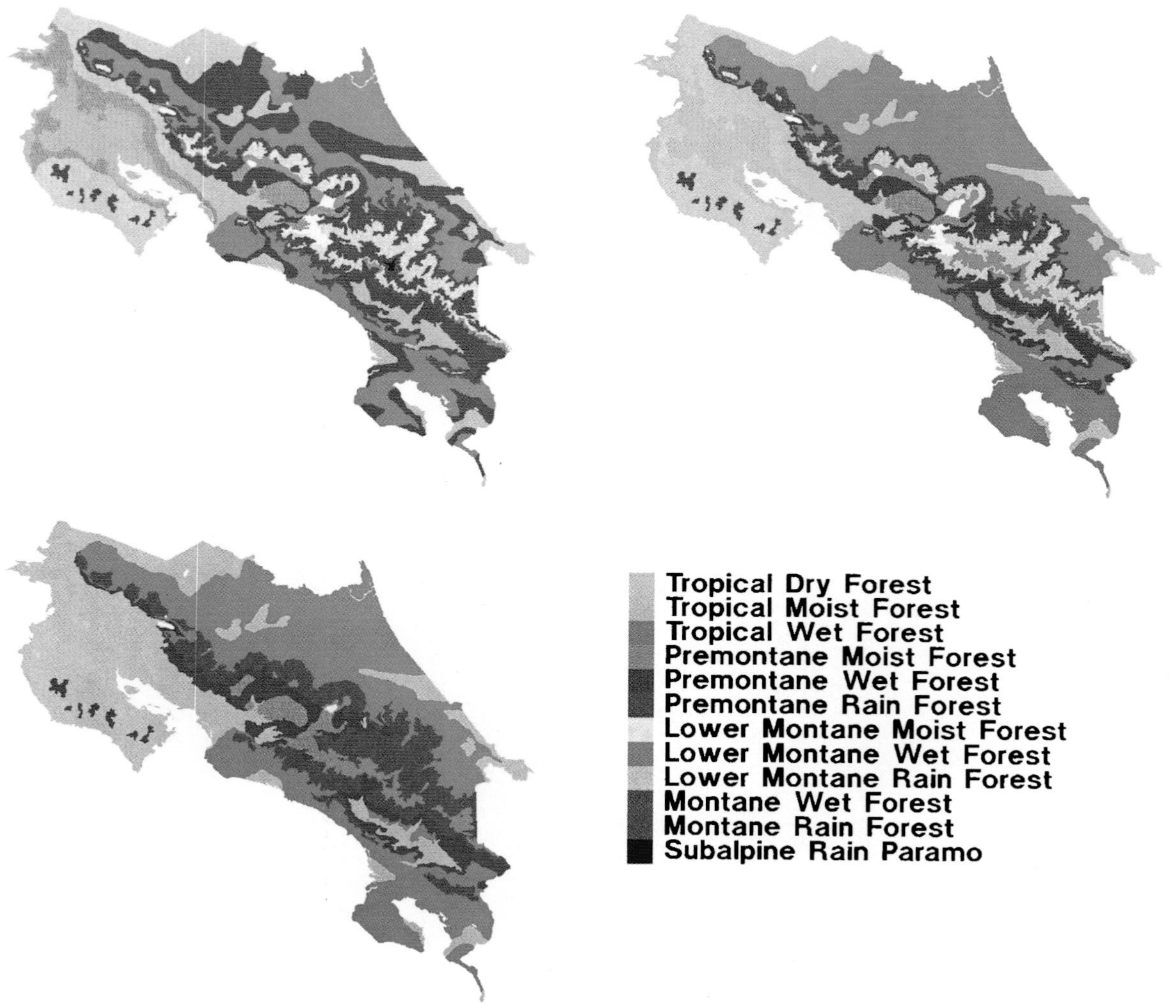

Plate 13 Provisional Holdridge life zones for Costa Rica under a) current climate, b) + 2.5°C climate change scenario, and c) + 3.6°C climate

Tropical Dry Forest
Tropical Moist Forest
Tropical Wet Forest
Premontane Moist Forest
Premontane Wet Forest
Premontane Rain Forest
Lower Montane Moist Forest
Lower Montane Wet Forest
Lower Montane Rain Forest
Montane Wet Forest
Montane Rain Forest
Subalpine Rain Paramo

7

The use of geographic information systems for analysis of scenery in the Cairngorm mountains, Scotland

David R. Miller, Jane G. Morrice, Paula L. Horne and Richard J. Aspinall

Introduction

Geographic information systems (GIS) offer a powerful tool for providing information to support decision-making in recreation planning and management, and to promote integrated management of resources based on their sensitivity to their use and the needs of local communities and visitors. This chapter describes an approach which uses GIS to assess resources from the perspective of recreation through evaluating scenery and visual impact. Existing GIS capabilities are also suited to analyzing the sensitivity of footpaths to erosion; to designing footpath networks, for example, by planning routes which avoid or visit particular features; to monitoring and helping to manage land use change, for example by managing the harvest of forest resources to minimize visual impacts; and to producing and evaluating land use scenarios, for instance in choosing optimal locations for visitor centres and other facilities.

A wider dimension to the approach presented in this chapter is given by legislation associated with European Community Directive 85/337 (European Communities, 1985). This requires that an environmental assessment be undertaken as part of a development plan or proposal, with the aim of a systematic approach providing 'relevant information on the environment concerned and the likely impact' (European Communities, 1989). Although attributes of the environment requiring study, including visual impact, are specified, assessment protocols are not specified. The GIS-based approaches used here provide protocols for assessing visual impact and may find wider application in environmental assessment. The integrative capacities of GIS allow diverse issues to be examined in an integrated manner, as information representing different perspectives and information bases is brought together in one system. This can assist in the discussion of management and planning questions (Aspinall, 1990).

The area considered in this chapter is Badenoch and Strathspey District

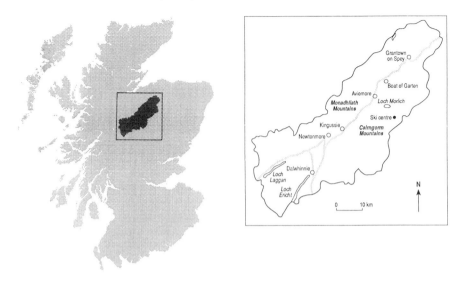

Figure 7.1   Location map of Badenoch and Strathspey District study area.

in the Highland Region of Scotland (Figure 7.1). Over 50 per cent of this District lies above 450 m, with a further 37 per cent between 250 and 420 m. The District also includes the Cairngorm mountains, the most extensive area of continuous high elevation in northwest Europe, with four peaks being over 1200 m. The Cairngorms are internationally recognized as one of the most important mountain areas in Britain (Ratcliffe, 1977; Curry-Lindahl *et al.*, 1982), containing subarctic, boreal, and moorland plant and animal communities (Curry-Lindahl, 1974; Conroy *et al.*, 1990) and unique wilderness landscapes (Watson, 1990). The combination of the range in altitude and the open panoramas associated with distinctive mountain topography contributes, in part, to the scenic appeal of the Cairngorm mountains. Landscapes and scenery are among the principal resources attracting tourists and other visitors to Badenoch and Strathspey District (CCS, 1990).

Recreation and tourism are important components of the local economy (Payne, 1972; Watson, 1988). In 1989, 1.9 million bed nights were recorded in the Spey Valley; many more visitors come for day trips, as the area is within three hours driving time of the population centres of Glasgow, Edinburgh, Dundee, and Aberdeen. Three studies particularly emphasize the importance of the scenery and landscape of Badenoch and Strathspey District for recreation and tourism. A total of 67 per cent of a sample of visitors to the District rated scenery as the most attractive/enjoyable visitor attraction (Watson, 1988), while 'high quality' scenery was rated as important/very important by 95 per cent of respondents in a questionnaire survey of visitors to the Cairngorms (MacKay Consultants, 1988). An earlier study identified drives, picnics and outings as the recreational activity with the greatest participation (59 per cent of visitors) (Survey Research Associates, 1981). The Cairngorms are also popular for a wide range of recreational activities, including hill walking, nature study, downhill skiing, and mountain biking (Watson, 1988; CCS, 1990). The potential and value of scenic and nature conservation interest of the Cairngorms is evidenced by a wide variety of national designations, such as National Scenic Area, National Nature Reserve, and Site of Special Scientific Interest. In many ways, it is equivalent to many of the National Parks of England and Wales, but there are no National Parks in Scotland.

While the qualities of the scenic and environmental heritage in the Cairngorms encourage tourism and recreation, they are highly sensitive to many of the resulting activities. Both a report by the Countryside Commission for Scotland (1990) and, more recently, the Cairngorms Working Party, which has been considering the management of the region, have identified a number of issues of direct relevance for recreation. These include the lack of a coherent strategy for the provision of facilities for visitors, the development of skiing facilities, an increase in footpath erosion, and the attrition of remote and wild mountain areas which are slow to recover from damage. Conservation, scenic, and recreational interests sit alongside traditional land uses. These different interests have frequently been in conflict, particularly when national and international interests are set against those of local communities; solutions to land management issues in the area must be sought through consensus and compromise (Buchanan-Smith, 1990).

GIS database for Badenoch and Strathspey District

The GIS which has been developed for Badenoch and Strathspey District incorporates data describing environmental and cultural resources, e.g. soils, land cover, land capability assessments for forestry and agriculture, nature conservation and landscape designations, visitor centres, roads, built-up areas. The database was compiled from maps and aerial photos and includes a digital terrain model (DTM) derived from the 1:250 000 scale, 100 m resolution data of the Directorate of Military Survey (Smith *et al.*, 1989). Land cover has been derived from the interpretation of LANDSAT satellite Thematic Mapper (TM) imagery, aerial photographs (Aspinall *et al.*, 1991; Gauld *et al.*, 1991; Miller *et al.*, 1991) and video photography.

Both raster and vector formats are used for the storage and manipulation of the data sets. Resolution of the raster data is between 25 m (for land cover) and 100 m; the variable quality of the input data sets is a prime concern and error analyses are integral to many of the analyses. Additionally, since the Cairngorm Visitor Survey (Mackay Consultants, 1988) found that 53 per cent of visitors had consulted Ordnance Survey (OS) maps as a source of information, and 42 per cent had consulted maps from another source, viewpoints designated on OS 1:50 000 Landranger maps are included in the

GIS. The area also has considerable forestry and, since tree type has an influence on visual assessment of landscape (Crowe, 1978), woodland distribution and type are included in the database.

Analysis of scenery using GIS

There are few methods for the objective analysis of scenery, even though this resource increasingly requires management. Rice (1988) divides Badenoch and Strathspey District into seven landscape tracts. Five of these occupy about 50 per cent of the District and comprise the higher land around the Spey catchment – the Cromdale Hills, Cairngorms, Badenoch (Dalwhinnie, Loch Ericht and south to Drumochter Pass), the Monadhliaths, and a northern moorland area from the Slochd to Lochindorb. These areas have peat, peaty gley, alpine and subalpine soils (Walker *et al.*, 1982) and associated moorland and blanket bog communities; mountain heath communities occur on the highest ground. These open vegetation communities contribute both to the high nature conservation value of the area (Ratcliffe, 1977) and to the scenic appeal through their wild land qualities (CCS, 1990).

Land at lower elevations between Glen Truim in the south and Grantown-on-Spey in the north also has a high scenic value; this is described as the middle valley landscape tracts (Rice, 1988). This division of the District, although useful for structuring discussion, says little of the specific merits and scenery of individual locations. GIS can be used to adopt the general principle of landscape classification while retaining attributes of the unique character of each individual location. Thus, GIS can provide a flexible framework for discussing the general context of scenic resources in the area and the contribution of each particular place.

In the same manner that locations may be classified by altitude, land cover, or land use, they may also be classified in terms of the amount of land visible from them; i.e., areas can be compared according to the relative area of land or the features visible from a location. The methods for such analysis can weight different components of a scene in a number of ways to emphasize particular components or features of interest. The results of modelling can be used for inventory, evaluation, or prediction (Aspinall, 1990) and provide information which can be used in the decision making. For example, the visual impact of a land cover or land use change can be measured in terms of the area from which it will be visible, or the impacts can be visualized in two- or three-dimensional or dynamic changes using simulation and computer display.

Three types of analysis are presented. Two relate to analysis of a large area (the whole of Badenoch and Strathspey District in this example) while one is more local and focuses on visual impact at a particular viewpoint in a particular direction. All are implemented within a GIS for Cairngorm mountains and Badenoch and Strathspey District.

Local analysis of scenery

GIS allows detailed studies of visual impacts at particular locations and for particular scenes. The scene viewed from a location can be divided into zones ranked according to a model of visual impact on an observer at that location, with respect to contributing environmental features such as topography, landform, and land cover. By applying principles of scene assessment (Land Use Consultants, 1990) and knowledge of the factors that influence object clarity, a model can be constructed, calibrated, and applied to describe the view from a specified location. Image rendering, with attenuation by atmospheric effects, is used to assess the visual clarity of objects in a landscape (McLaren and Kennie, 1989).

Three further factors are considered in the model.

1. Distance depth cueing and size perspective. A distance-decay function is employed to calculate visual clarity as the distance between an observer and an object increases. A projective transformation is also employed. This calculates a score which incorporates the effect of an object's physical dimensions.
2. Angle of intersection of the view with the terrain. As the angle of view increases with respect to the normal of the terrain, the proportion of terrain visible from a location decreases. However, if land cover at a point has definite height (for example a forest or buildings) the angle will decrease and visual impact will increase. This effect will be most apparent on the visible horizon or for tall objects.
3. Land cover. Land cover types within a scene provide the context against which to analyse and model visual impact. Texture, colour, hue, and brightness are attributes of land cover type. Their role is also influenced by distance and atmospheric effects. In general, the discrimination of objects and their contrast against the background diminishes with distance. Significant visual impacts occur where contrasts in hue or brightness between background land cover and the object being viewed are greatest. For example, although a land cover change may occur and be visible from a viewpoint, it need not be discernible from the background; the visual impact of that change should be identified as low by the model. Texture, colour and intensity for land cover are derived from LANDSAT TM satellite imagery in the GIS and from digital analysis of video photographs of land cover types in the study area.

The model used to describe the scene from a viewpoint produces a simple index (V_i) and has been applied in the GIS. The model is:

$$V_i = 1/D^2 + bH + Lf - Av \qquad (7.1)$$

where D = distance from viewpoint, H = distance below horizon, b = a weighting factor as an attenuation function for distance below the horizon, Lf = weighting according to landform type, Av = vertical angle to feature.

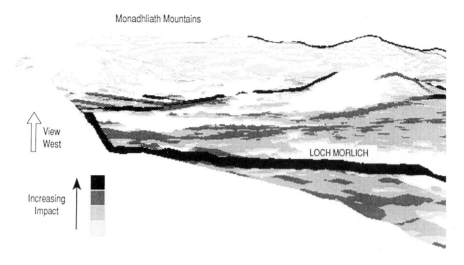

Figure 7.2   Impact model of forest land use from the viewpoint in the car park at Cairngorm ski area.

The viewpoint from the car park at the Cairngorm Ski Centre towards Aviemore is presented as an example of this local analysis of scenery (Figure 7.2). Forest cover is recoded according to average heights for tree type in the area. The addition of these data to the topographic background provides an information framework against which to measure the impact of a change in forest land use on the view from the viewpoint. A true perspective view from the viewpoint, along a bearing of 270° (toward Aviemore) is derived from the digital terrain data and rendered with the output from the model of scenic impact. The use of this output to visualize and measure the impact of forest land use change in the example scene shows that the visual impact of afforestation is generally low, being below the horizons and blending with the dark texture background of the mountains.

Wide area analysis

Analysis of scenery from an individual viewpoint is of limited use in a strategic planning or management context although it serves as an effective *aide memoire* to supplement site visits (Davidson and Selman, personal communication) and can be a useful tool for removing cloud and mist. GIS can also be used to investigate scenery across large areas, serving to compare particular locations in a wider geographic context. This synoptic analysis is a particular strength of GIS, and typically is a more appropriate use of environmental data which are often of variable quality and scale.

The DTM for Badenoch and Strathspey District is used to identify land visibility using viewshed and intervisibility functions in GIS (Fisher, 1992; 1993). This analysis is used as the basis for assessment of potential visual impact throughout the District. Two approaches are used.

1. Calculation of viewsheds for selected visitor viewpoints, to identify priority zones of visual importance to tourists. An intervisibility calculation is
   made for selected locations, and the map output is recoded to describe the
   number of viewpoints that can 'see' each location in the District. The
   application of this approach in architectural design is described by Benedikt
   (1979), who describes the set of all locations visible from any specified
   location as the 'isovist.' He notes the central importance of the observer's
   location because it is 'representing the position of the observer whose
   spatial experience we are trying to explore.' The objective identification of
   the boundaries of this spatial experience is the basis for this analysis of the
   digital terrain data. Fisher (1992, 1993) has described an extension of
   viewshed analysis which identifies uncertain, or fuzzy, viewshed boundaries and may be used to account for uncertainties associated with digital
   terrain data.
2. A census of the total area visible from all locations within the District
   (each location is a pixel in the raster data set). This describes the relative
   (and absolute) intervisibility of land within the District. An intervisibility
   calculation is made for every pixel in the digital terrain model, recording
   the total number of pixels visible from each. This is a district-wide version
   of the previous analysis.

These intervisibility analyses both use the equation:

$$Nv = \sum_{1°}^{360°} \left(\sum_{0.1\,km}^{100\,km} (x_k > mx_{k-1}) \right) \tag{7.2}$$

where

$$Nv = \text{number of visible pixels (visible area to observer)}$$
$$x_k = \text{height}_k + \text{earth curvature correction} +$$
$$\text{atmospheric refraction correction}$$
$$mx_{k-1} = \text{maximum height of } x_1 \to x_{k-1}$$

Land cover effects may be included in this analysis by adjusting the
height of each location. The equation may be applied to individual locations
or to all locations in the area. Vertical angles are calculated making correc-
tions for earth curvature and atmospheric refraction. The analysis considers
a complete 360° rotation around each location for a radius of up to 100 km
and is carried out on the 100 m resolution raster DTM.

Analysis for viewpoints

The analysis of scenery by visitors derives from consideration of the most
popular recreational activity, drives, picnics and outings, and the viewpoints
to which these visitors may be attracted. The locations used as centres for the
calculation of intervisibility are the five viewpoints identified on the OS

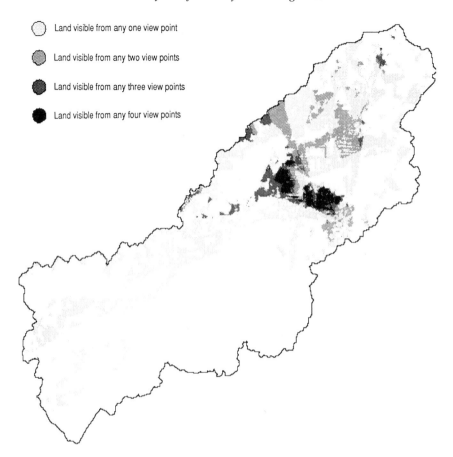

○ Land visible from any one view point

◍ Land visible from any two view points

◕ Land visible from any three view points

● Land visible from any four view points

Figure 7.3   Land visibility maps for Badenoch and Strathspey District: terrain ranked by visibility from the five viewpoints indicated on OS Landranger maps.

Landranger maps for Badenoch and Strathspey District. Equation (7.1) is applied to each of the viewpoints in turn. Scenic potential is then measured by determining the number of viewpoints from which any terrain location is visible. Figure 7.3 shows the District zoned according to the visibility of land from the five viewpoints. Forest cover, which may intrude on the scene or obscure the view, is included in this analysis. Scores range from 0 (land not visible from any of the viewpoints) to 4 (land visible from any four view-points). No location in the District is visible from all five viewpoints.

Analysis of Badenoch and Strathspey District

For the census approach equation (7.2) is applied to all locations in the District. Land outside the District which is visible from within it is included to ensure

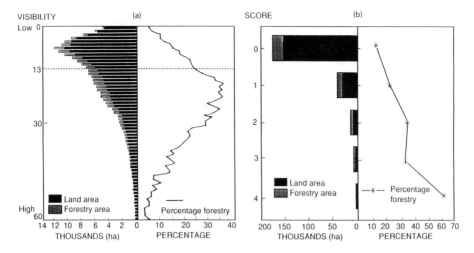

Figure 7.4 (a) Land and forestry area for the census data set, divided into bands, each of 610 ha of visible land (b) Land and forestry area scored according to the number of viewpoints from which it is visible.

that results are not influenced by its administrative boundary; forest cover is not included although this may obscure views and impact on intervisibility. The calculation produces the total area of land visible from every location (pixel). The maximum area visible from any individual location in this example is 1555·5 km^2 and the mode is 42·7 km^2.

Figures 7.4 and 7.5 illustrate the results of this analysis. In Figure 7.4a, the visibility census data set is presented in units of 610 ha (the minimum area visible from any point in the District) to illustrate the pattern of relative area of visibility throughout the District. For diagrammatic purposes, the histogram of visible areas (ranging from 610 ha to 1555·5 km^2) (Figure 7.4b) has been used in Figure 7.5 to present the results as two equal-area groups representing high and low visibility. The higher-visibility land is along the bottom and sides of the Spey valley while, in general, the more mountainous regions of the District have lower values because of their relatively enclosed nature. The five upland landscape tracts identified by Rice (1988) broadly coincide with the low-visibility land. However, there are exceptions, notably the high visibility of the Slochd, and mountain summits such as Cairngorm and on certain hill and valley sides.

Application in planning

These visibility analyses demonstrate three approaches to the objective analysis of elements of landscape with respect to the impacts of land use on an observer. In particular, the assessments provide an audit or inventory of

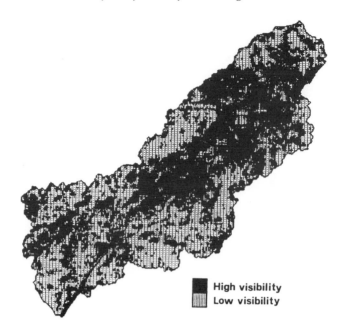

Figure 7.5   Land visibility maps for Badenoch and Strathspey District: terrain visibility census, divided into High and Low visibility by equal areas of visible land.

intervisibility in the District against which land cover, land use, and other activities and resources may be compared and complies with one objective of the European Communities' Environmental Assessment Directive. This type of wide area assessment highlights areas of greatest importance with respect to visitors across the District and is useful for strategic planning and management. The methods presented here allow an assessment of visual impacts which is sensitive to the geographic context within which they occur. Such methods provide a role for GIS in data synthesis, analysis, and evaluation as part of visual impact assessment. This type of analysis has several applications, including providing some basic tools which can contribute to environmental assessment.

Maps such as Figure 7.3 can be used to assess the relative impact of different land covers and uses on scenery, and GIS can be used to delimit areas which, in terms of potential visual impact, are most sensitive to land cover change. This provides a framework within which to assess the impact of any proposed changes in land use. The analysis can therefore be used to help plan the location of planting and harvesting programmes for forestry while considering the contribution to the scenery of the District. It can also be used to assess the contribution and importance of existing land cover features. For example, Figure 7.4b shows the area of land under different forest cover types visible from viewpoints (Figure 7.2). Forests, over 90 per cent of which are coniferous, occupy over 60 per cent of the land visible from four viewpoints. This proportion is 10 times that in the most widely visible land within the District and almost twice the proportion of forestry visible in the District. This illustrates the high impact of forestry on the visual landscape seen by a visitor. In particular, the visual landscape for visitors to viewpoints indicated on the OS Landranger maps appears to emphasize forestry as a land cover type.

The quantification of the area visible from any location is an objective measure of the extent to which a change in land cover will be visible to an observer. It also allows for comparison between different locations and creates a framework within which to assess relative visibility in an area. In Badenoch and Strathspey District, the measures of visual impact described illustrate that forest land cover has a high visual profile. This has implications for forest management and also for the management of landscapes which are of high scenic interest and are an important element in the local economy as visitors are attracted to the scenery.

At present, these GIS tools for the analysis of scenery are used for either historical evaluation of the impacts of land use change (Gauld *et al.*, 1991; Miller *et al.*, 1991) or prediction of the impacts of future changes (Aspinall, 1990). Future development of these methods, focusing on their application in regional and local land management and planning, will provide useful information for the assessment of the amenity potential of landscape and also support decision-making. The methods presented here can also be extended to other recreational groups and activities. For example, GIS can be used to generate views, from hill footpaths, minor roads, or laybys, which can be integrated to provide a more wide-ranging description of scenic interest and the sensitivity of resources to visitor pressure and land cover changes.

The approach can also be linked to more traditional methods of landscape analysis, such as the use of questionnaires to assess perception of place. Such a study has recently been initiated in the Cairngorms as a joint programme between the Macaulay Land Use Research Institute and the Department of Landscape Architecture at Heriot-Watt University, Edinburgh. This study is designed to elucidate constructs which underpin individual preference for particular landscapes and scenery. It may be linked to GIS through analysis of what may be termed 'non-georeferenced locational data' and used to enhance landscape evaluation, assessment, and interpretation through greater understanding of experience of landscape. This approach offers potential for analysis of the visual environment as perceived by different groups (e.g. tourists, residents) and offers possibilities for multiple representation of the same scene under a range of scenarios with respect to both observer and land use.

Acknowledgments

The authors would like to thank Alfred O'Dell at RAE Farnborough for development of software, and the Directorate of Military Survey for experimental use of their 1:250 000 digital terrain data. Thanks also go to Calder Miller of the Forestry Commission for information relating to Environmental Assessments. The work has been funded by the Scottish Office Agriculture and Fisheries Department as part of a programme of research into the development of GIS-based analytical methods for application in land use science.

References

Aspinall, R.J., 1990, An integrated approach to land evaluation: Grampian Region in Bibby, J.S. and Thomas, M.F. (Eds) *The Evaluation of Land Resources in Scotland*, Aberdeen: Macaulay Land Use Research Institute, pp. 71–80.

Aspinall, R.J., Miller, D.R. and Birnie, R.V., 1991, From data source to database: acquisition of land cover information for Scotland, in *Remote Sensing of the Environment. Proceedings of Image Processing 91*, Birmingham, 131–52.

Benedikt, M.L., 1979, To take hold of space: isovists and isovist fields, *Environment and Planning B*, **6**, 47– 65.

Buchanan-Smith, A., 1990, Foreword, in Conroy, J.W.H., Watson, A. and Gunson, A.R. (Eds) *Caring for the High Mountains – Conservation of the Cairngorms*, Aberdeen: Centre for Scottish Studies.

Conroy, J.W.H., Watson, A. and Gunson, A.R., 1990, *Caring for the High Mountains – Conservation of the Cairngorms*, Aberdeen: Centre for Scottish Studies.

Countryside Commission for Scotland (CCS), 1990, *The Mountain Areas of Scotland: Conservation and Management*, Battleby: Countryside Commission for Scotland.

Crowe, S., 1978, *The landscape of forests and woods*, Forestry Commission Booklet 44, London: HMSO.

Curry-Lindahl, K., 1974, *Survey of Northern and Western European National Parks and Equivalent Reserves*, Nairobi: United Nations Environment Programme.

Curry-Lindahl, K., Watson, A. and Watson, R.D., 1982, *The Future of the Cairngorms*, Aberdeen: North-east Mountain Trust.

European Communities, 1985, The assessment of the effects of certain public and private projects on the environment, *Official Journal of the European Communities* 85/337/EEC.

European Communities, 1989, *22nd General Report on the Activities of the European Communities 1988*, para 550, Luxembourg.

Fisher, P.F., 1992, First experiments in viewshed uncertainty: simulating the fuzzy viewshed, *Photogrammetric Engineering and Remote Sensing*, **58**, 345–52.

Fisher, P.F., 1993, Probable and Fuzzy Models of the Viewshed Operation, in *Proceedings of GISRUK'93*, Keele University.

Gauld, J.H., Bell, J.S., Towers, W. and Miller, D.R., 1991, The measurement and analysis of land cover changes in the Cairngorms, Report to the Scottish Office Environment Department and the Scottish Office Agriculture and Fisheries Department, Aberdeen: Macaulay Land Use Research Institute.

Land Use Consultants, 1990, Landscape Assessment: principles and practice, Report to the Countryside Commission for Scotland. Perth: Countryside Commission for Scotland.

MacKay Consultants, 1988, Cairngorm Visitor Survey: Summer 1987, Report for the Countryside Commission for Scotland, Highlands and Islands Development Board, Highland Regional Council and the Nature Conservancy Council, Perth: Countryside Commission for Scotland.

McLaren, R.A. and Kennie, T.J.M., 1989, Visualisation of digital terrain models: techniques and applications, in Raper, J. (Ed.), *Three dimensional applications in geographic information systems*, London: Taylor & Francis, 79–98.

Miller, D.R. Gauld, J.H., Bell, J.S. and Towers, W., 1991, Land Cover Change in the Cairngorms, in *Proceedings, AGI 1991*, 2.13.1–7.

Payne, M.A., 1972, *Speyside project, Second Interim Report*, Perth: Countryside Commission for Scotland.

Ratcliffe, D.A., 1977, *A Nature Conservation Review*, Cambridge: Cambridge University Press.

Rice, D.E.A., 1988, The landscape of the Spey catchment, in Jenkins, D. (Ed.), *Land use in the River Spey catchment*, ACLU Symposium 1, Aberdeen: Aberdeen Centre for Land Use.

Smith, J.M., Miller, D.R. and Morrice, J.G., 1989, An evaluation of a low resolution Digital Terrain Model with satellite imagery for environmental mapping and analysis, in *Remote Sensing for Operational Applications*, Remote Sensing Society Annual Conference, Bristol.

Survey Research Associates, 1981, *Market survey of holidaymaking and countryside recreation in Scotland*, Edinburgh: Scottish Tourist Board.

Walker, A.D., Campbell, C.G.B., Heslop, R.E.F., Gauld, J.H., Laing, D., Shipley, B.M. and Wright, G.G., 1982, *Eastern Scotland*, Soil and Land Capability Handbook 5, Aberdeen: Macaulay Institute for Soil Research.

Watson, R.D., 1988, Amenity and informal recreation on and around the River Spey: a consideration of conflicts, issues and possible solutions, in Jenkins, D. (Ed.) *Land use in the River Spey catchment*, ACLU Symposium 1, Aberdeen: Aberdeen Centre for Land Use.

Watson, R.D., 1990, A hillman looks at the conservation of the Cairngorms, in Conroy, J.W.H., Watson, A. and Gunson, A.R. (Eds) *Caring for the High Mountains – Conservation of the Cairngorms*, Aberdeen: Centre for Scottish Studies, 91–107.

PART II

Evaluation of Natural Hazards

8

GIS in landslide hazard zonation: a review, with examples from the Andes of Colombia

Cees J. van Westen

Introduction

Mass movements in mountainous terrain are natural degradational processes, and one of the most important landscape building factors. Most of the terrain in mountainous areas has been subjected to slope failure at least once, under the influence of a variety of causal factors, and triggered by events such as earthquakes or extreme rainfall.

Mass movements become a problem when they interfere with human activity. The frequency and the magnitude of slope failures can increase due to human activities, such as deforestation or urban expansion (Figures 8.1 and 8.2). In developing countries, this problem is especially great due to rapid non-sustainable development of natural resources. Developing countries suffer some 95 per cent of total disaster-related fatalities, which are estimated to number on the order of 225 000 per year (Hansen, 1984). Economic losses attributable to natural hazards in developing countries may represent as much as 1–2 per cent of gross national product (Fournier D'Albe, 1976). Losses due to mass movements are estimated to be one quarter of the total losses due to natural hazards (Hansen, 1984). These statistics illustrate well the importance of hazard mitigation. Indeed, the decade 1990–2000 has been designated the International Decade for Natural Disaster Reduction by the General Assembly of the United Nations.

Mitigation of landslide disasters can be successful only when detailed knowledge is obtained about the expected frequency, character, and magnitude of mass movement in an area. The zonation of landslide hazard must be the basis for any landslide mitigation project and should supply planners and decision-makers with adequate and understandable information. Analysis of landslide hazard is a complex task, as many factors can play a role in the occurrence of mass movements (see Crozier, 1986 for a comprehensive treatment of causes). The analysis requires a large number of input parameters, and techniques of analysis may be very costly and time-consuming. The

Figure 8.1   Landslide in the Rio Chinchina catchment, Colombia.

Figure 8.2   Landslide in the city of Manizales, Colombia.

increasing availability of computers during the last decades has created opportunities for a more detailed and rapid analysis of landslide hazard. This chapter describes in a general way the results of an international research project dealing with the application of GIS in landslide hazard zonation. For more information the reader is referred to Soeters *et al.* (1991), Rengers (1992), Rengers *et al.* (1992), Van Asch *et al.* (1992), van Westen and Alzate (1990) and van Westen (1992a, 1992b, 1993).

Definitions

Mass movement is defined as 'the outward or downward gravitational movement of earth material without the aid of running water as a transporting agent' (Crozier, 1986). Although by definition the term landslide is used only for mass movements occurring along a well-defined sliding surface, it has been used as the most general term for all mass movements, including those that involve little or no sliding. In this study the terms mass movement; landslide; slope movement; and slope failure are used synonymously.

To differentiate between the terms hazard; vulnerability; and risk, the following definitions (given by Varnes, 1984) have become generally accepted:

Natural hazard (H):   the probability of occurrence of a potentially damaging phenomenon within a specified period of time and within a given area.

Vulnerability (V):   the degree of loss to a given element or set of elements at risk (see below) resulting from the occurrence of a natural phenomenon of a given magnitude. It is expressed on a scale from 0 (no damage) to 1 (total loss).

Specific risk (Rs):   the expected degree of loss due to a particular natural phenomenon. It may be expressed by the product of H and V.

Elements at risk (E):   the population, properties, economic activities, including public services, etc. at risk in a given area.

Total risk (Rt):   the expected number of lives lost, persons injured, damage to property, or disruption of economic activity due to a particular natural phenomenon. It is therefore the product of specific risk (Rs) and elements at risk (E):

$$Rt = (E) * (Rs) = (E) * (H*V).$$

Landslide hazard is commonly shown on maps, which display the spatial distribution of hazard classes (landslide hazard zonation). Zonation refers to 'the division of the land in homogeneous areas or domains and their ranking according to degrees of actual/potential hazard caused by mass movement' (Varnes, 1984). Landslide hazard zonation requires a detailed knowledge of the processes that are or have been active in an area, and of the factors leading to the occurrence of the potentially damaging phenomenon. This is considered the task of earth scientists. Vulnerability analysis requires detailed knowledge of the population density, infrastructure, and economic activities, in addition to the hazard. Therefore, this part of the analysis is done mainly by persons from other disciplines, such as urban planning, social geography, and economics.

Fully-developed examples of risk analysis on a quantitative basis are still scarce in the literature (Einstein, 1988), because of the difficulties in defining quantitatively both hazard and vulnerability. Hazard analysis is seldom executed in accordance with the definition given above, since the probability of occurrence of potentially damaging phenomena is extremely difficult to determine for larger areas. The determination of actual probabilities requires analysis of triggering factors, such as earthquakes or rainfall, or the application of complex models. In most cases, however, there is no clear relationship between these factors and the occurrence of landslides. Therefore, the legend classes used in most hazard maps do not give more information than relative indications, such as high, medium, and low hazard. This study is restricted to the analysis of landslide hazard.

Trends in landslide hazard zonation

A large amount of research on hazard zonation has been done over the last 30 years, as the consequence of an urgent demand for slope instability hazard mapping. Overviews of the various slope instability hazard zonation techniques can be found in Hansen (1984), Varnes (1984), and Hartlén and Viberg (1988). The initial investigations were oriented mainly toward problem solving at the scale of site investigation and development of deterministic models. A wide variety of deterministic slope stability methods is now available to the engineer (Chowdury, 1978; Graham, 1984).

The large regional variability of geotechnical variables, such as cohesion; angle of internal friction; thickness of layers; or depth to groundwater, is inconsistent with the homogeneity of data required in deterministic models. The site investigation approach provides an unacceptable cost/benefit ratio for engineering projects over larger areas during the planning and decision-making phases due to the high cost and time requirements of data collection. Several types of landslide hazard zonation techniques have been developed to tackle such problems encountered in the application of deterministic modelling. A summary of the various trends in the development of techniques is given in Table 8.1. Each of the main groups highlighted in Table 8.1 is described in more detail in the following paragraphs.

Previous landslide studies using GIS

The development of GIS has greatly increased the availability of techniques for landslide hazard assessment and their application. The first applications of simple, self-programmed, prototype GIS in the analysis of landslide hazard zonation date from the late 1970s (Burrough, 1986). Newman *et al.* (1978) reported on the feasibility of producing landslide susceptibility maps using computers. Carrara *et al.* (1978) reported results of multivariate analysis applied

Table 8.1   General trends in landslide hazard methods

Type of landslide hazard analysis	Main characteristic
A. Distribution analysis	Direct mapping of mass movement features resulting in a map which gives information only for those sites where landslides have occurred in the past
B. Qualitative analysis	Direct, or semi-direct, methods in which the geomorphological map is renumbered to a hazard map, or in which several maps are combined into one using subjective decision rules, based on the experience of the earth scientist
C. Statistical analysis	Indirect methods in which statistical analyses are used to obtain predictions of the mass movement hazard from a number of parameter maps
D. Deterministic analysis	Indirect methods in which parameter maps are combined in slope stability calculations
E. Landslide frequency analysis	Indirect methods in which earthquake and/or rainfall records or hydrological models are used for correlation with known landslide dates, to obtain threshold values with a certain frequency

on grid cells with a ground resolution of 200 m × 200 m using approximately 25 variables. Huma and Radulescu (1978) reported an example from Romania of a qualitative hazard analysis including the factors of mass movement occurrence, geology, structural geological conditions, hydrological conditions, vegetation, slope angle, and slope aspect. Radbruch-Hall *et al.* (1979) wrote their own software to produce small-scale (1:7500 000) maps with 6 million pixels showing hazards, unfavourable geological conditions, and areas where construction or land development may exacerbate existing hazards. The maps were made by qualitative overlay of several input maps.

During the 1980s, the use of GIS for slope instability mapping increased sharply due to the development of commercial GIS systems, such as ARC/INFO, Intergraph, SPANS, ILWIS and IDRISI, and the increasing availability of personal computers (PCs). The majority of case studies presented in the literature on this subject deal with qualitative hazard mapping. The importance of geomorphological input data is stressed in the methods used by Kienholz *et al.* (1988), who used a GIS for a qualitative mountain hazard analysis; detailed aerial photo interpretation was used as a basis. The authors state that, due to the lack of good models and geotechnical input data, the use of a relatively simple model based on geomorphology seems to be the most realistic method. Most examples of qualitative hazard analysis with GIS are very recent (Stakenborg, 1986; Mani and Gerber, 1992; Kingsbury *et al.*, 1992). Many examples are presented in the proceedings of the First International Symposium on the use of Remote Sensing and GIS for Natural Risk Assessment, held in Bogotá, Colombia, in March 1992 (Alzate, 1992).

Examples of landslide susceptibility analysis with GIS reported since the 1970s have come mainly from the United States Geological Survey (USGS) in Menlo Park, California, where Brabb and his team have proceeded with their work and extended it, taking into account additional factors besides landslides, geology, and slope (Brabb, 1984; Brabb *et al.*, 1989). Other examples of quantitative univariate statistical analysis with GIS are rather scarce (Lopez and Zinck, 1991; Choubey and Litoria, 1990). This is rather strange, since one of the strong advantages of using a GIS is the capability to test the importance of each factor, or combinations of factors, and assign quantitative weighting values based on mass movement density.

Recent examples of multivariate statistical analysis using GIS have been presented mainly by Carrara and his team. Their work has developed from the use of large rectangular grid cells as the basis for analysis (Carrara *et al.*, 1978; Carrara, 1983, 1988) towards the use of morphometric units (Carrara *et al.*, 1990, 1991). The method itself has not undergone major changes. The statistical model is built-up in a training area, where the spatial distribution of landslides is (or should be) well known (Carrara, 1988). In the next step the model is extended to the whole study area or target area, based on the assumption that the factors that cause slope failure in the training area are the same as in the target area.

Another example of multivariate analysis using a GIS is presented by Bernknopf *et al.* (1988). They applied multiple regression analysis to a data set, using presence/absence of landslides as the dependent variable and the factors used in a slope stability model (soil depth, soil strength, slope angle) as independent variables. Water table data and cohesion data were not taken into account. The resulting regression function is transformed so that landslide probability can be calculated for each pixel.

Deterministic modelling of landslide hazard using GIS has become rather popular. Most examples deal with infinite slope models, since they are simple to use for each pixel separately (Brass *et al.*, 1989; Murphy and Vita-Finzi, 1991). Hammond *et al.* (1992) presented methods in which the variability of the factor of safety is calculated from selected input variables following the Monte Carlo technique. This implies a large number of repeated calculations, which require the use of a GIS.

A relatively new development in the use of GIS for slope instability assessment is the application of so-called neighbourhood analysis. Most of the conventional GIS techniques are based on map overlaying, which allows only for the comparison of different maps at the same pixel locations. Neighbourhood operations also allow evaluation of the neighbouring pixels around a central pixel, and can be used in the automatic extraction from a digital terrain model (DTM) of such morphometric and hydrological features as slope angle; slope aspect; downslope and cross-slope convexity; ridge and valley lines; catchment area; stream ordering, and the contributing area for each pixel. An overview of the algorithms applied in the extraction of morphometric parameters from GTMs is given by Gardner *et al.* (1990). Carrara identified automatically the homogeneous units he used as the basis for a multivariate analysis from a detailed DTM. The morphometric and hydrological parameters used in that analysis were also extracted automatically (Carrara *et al.*, 1990). Niemann and Howes (1991) performed a statistical analysis based on automatically extracted morphometric parameters (slope angle, slope aspect, down-slope and cross-slope convexity and drainage area), which they grouped into homogeneous units using cluster analysis. Van Dijke and van Westen (1990) applied a simple type of neighbourhood analysis to model the runout distances for rockfall blocks.

A recent development in the use of GIS for slope instability zonation is the application of expert systems. Pearson *et al.* (1991) developed an expert system in connection with a GIS in order to 'remove the constraints that the users should have a considerable experience with GIS'. A prototype interface between a GIS (ARC/INFO) and an expert system (Nexpert Object) was developed and applied for translational landslide hazard mapping in an area in Cyprus. The question remains, however, whether the rules used in the expert system apply only to this specific area, or whether they are universally applicable.

Pilot study areas

Selecting the working scale for a slope instability analysis project is determined by the purpose for which it is executed. The following scales of analysis, which were presented in the International Association of Engineering Geologists' monograph on engineering geological mapping (IAEG, 1976), can also be distinguished in landslide hazard zonation:

- Synoptic or regional scale (<1:100 000);
- Medium scale (1:25 000–1:50 000);
- Large scale (1:5000–1:10 000).

Regional-scale hazard analysis is used to outline problem areas with potential slope instability. The maps are mainly intended for agencies dealing with regional (agricultural, urban, or infrastructural) planning. The areas to be investigated are very large, on the order of 1000 km² or more, and the required detail of the map is low. The maps indicate regions where severe mass movement problems can be expected to threaten rural, urban, or infrastructural projects. Terrain units with areas of at least several tens of hectares are outlined. Within these units the degree of hazard is assumed to be uniform.

Medium-scale hazard maps are made mainly for agencies dealing with inter-municipal planning or companies dealing with feasibility studies for large engineering works (such as dams, roads, railroads). The areas to be investigated will have areas of several hundreds of square kilometres. At this scale considerably more detail is required than at the regional scale. The maps may serve, for example, for the choice of corridors for infrastructural construction or zones for urban development. The detail should be such that adjacent slopes with the same lithology are evaluated separately, which may result in different hazard scores, depending on other characteristics, such as slope angle and land use. Even within a single terrain unit, a distinction should be made between different slope segments, for example a concave part of a slope should receive a different score than an adjacent straight slope.

Large-scale hazard maps are produced mainly for authorities dealing with detailed planning of infrastructural, housing, or industrial projects, or with evaluation of risk within a city. The size of an area under study would be on the order of several tens of square kilometres. The hazard classes on such maps should be absolute, indicating, for example, the probability of failure for each individual unit with an area down to less than a hectare.

Although the selection of the scale of analysis is usually determined by the intended application of its results, the choice of technique for mapping landslide hazard remains open. The choice depends on the type of problem, the availability of geotechnical and other data, the availability of financial resources, and time restrictions, as well as on the knowledge and experience of the research team.

Three pilot study areas for the different working scales, defined above, were selected in the Rio Chinchina catchment in Colombia. The catchment of the Rio Chinchina, with a surface area of 722 km^2 and a perimeter of 159 km, is located on the western slope of the central Andean mountain range (Cordillera Central) in the Caldas Department in Colombia (see Figure 8.3). This area was chosen as the study area to test the methodology developed in this work because of its following characteristics.

1. The severity of natural hazards in the area, combined with intensive industrial and agricultural activity and a high population density, has caused considerable damage and loss of lives in the past. The area is susceptible to mass movement, earthquake, and volcanic hazards. The largest disaster in the Rio Chinchina area took place on 13 November 1985, when a lahar, triggered by an eruption of the glacier-covered Nevado del Ruiz, caused the death of about 2000 persons and destroyed all bridges over the Rio Claro and Rio Chinchina. The last major earthquake, which killed 50 persons in Pereira and Manizales and caused considerable property damage, occurred on 23 November 1979. Landslide casualties and material damage are reported almost annually, in both urban and rural areas. The road network also suffers from severe mass movement problems. The so called *triangulo vial* (road triangle) between Manizales, La Manuela, and

Figure 8.3   The pilot study areas for the three working scales, located in the Rio Chinchina catchment, central Colombia.

Chinchina is considered by the Ministry of Public Works to be one of the main problem areas in the Colombian road network (Baez *et al.*, 1988).
2. The availability of maps, aerial photos, and reports. Imagery and topographic maps, as well as a wide range of thematic maps from different years and at different scales, are available for most parts of the area.

Input data

A list of the various input data needed to assess landslide hazard at the regional, medium, and large scale is given in Table 8.2. The list is extensive, and only in an ideal case will all types of data be available. However, the amount and type of data that can be collected, determine the type of hazard analysis that can be applied, ranging from qualitative assessment to complex statistical methods.

The data layers needed to analyse landslide hazard can be subdivided into five main groups: geomorphological; topographic; engineering geological or geotechnical; land use; and hydrological data. A data layer in a GIS can be seen as one digital map, containing one type of data, composed of one

Table 8.2 Overview of input data needed for landslide hazard analysis

Data types	Summary of data collection techniques	Feasibility of data collection		
		Regional scale	Medium scale	Large scale
GEOMORPHOLOGY				
1. Terrain mapping units (TMUs)	Satellite image interpretation + walk over study	High	Moderate	Low
2. Geomorphological units	Aerial photo interpretation and field check	Moderate	High	High
3. Geomorphological subunits	Aerial photo interpretation and field check	Low	High	High
4. Landslides (recent)	Aerial photo interpretation and field descriptions	Low	High	High
5. Landslides (older period)	Aerial photo interpretation + collection of landslide records from newspapers, fire brigades, or church archives	Low	High	High
TOPOGRAPHY				
6. Digital terrain model (DTM)	Collection of existing contour maps	Moderate	High	High
7. Slope map (degrees or percentage)	Made from a (DTM)	Moderate	High	High
8. Slope direction map	Made from a DTM, no extra data collection required	Moderate	High	High
9. Breaks of slope	Aerial photo interpretation	Low	Moderate	High
10. Concavities/convexities	Made from a DTM, or detailed photo interpretation	Low	Low	High
ENGINEERING GEOLOGY				
11. Lithologies	Checking of existing geological maps, or by mapping if no data are available	Moderate	High	High
12. Material sequences	Made by a combination of other maps (geomorphological, geology, slope and DTM)	Low	Moderate	High
13. Sampling points	Field descriptions of soil and rock outcrops, and laboratory analysis of selected samples to characterize material types	Moderate	High	High

Item	Method			
14. Faults & lineaments	Satellite image interpretation, aerial photo interpretation, and fieldwork	High	High	High
15. Seismic events	Collection of existing seismic records	High	High	High
16. Isolines of seismic intensity	Questionnaires on the observed damage from earthquake(s)	Low	Moderate	High
LAND USE				
17. Infrastructure (recent)	Aerial photo interpretation and topographic map	Moderate	High	High
18. Infrastructure (older)	Aerial photo interpretation and topographic map	High	High	High
19. Land use map (recent)	Aerial photo interpretation and classification of satellite images and field check	Moderate	High	High
20. Land use map (older)	Aerial photo interpretation	Moderate	High	High
21. Cadastral blocks	Collection of existing cadastral maps and database	Low	Low	High
HYDROLOGY				
22. Drainage	Aerial photo interpretation and topographic map	High	High	High
23. Catchment areas	Aerial photo interpretation and topographic map or modelling from a DTM	Moderate	High	High
24. Meteorological data	Collection of existing meteorological stations	High	High	High
25. Water table	Field measurements of Ksat and modelling	Low	Low	Moderate

type of element (points, lines, units), and having one or more accompanying tables. Of course, the layers that have to be taken into account vary for different environments. Tectonic data, for example, are not needed in an area that is seismically inactive, and in some areas it may be necessary to include types of data not listed in the Table.

The second column of Table 8.2 gives a summary of the method by which each data layer is collected, referring to the three phases of data collection (image interpretation, fieldwork, and laboratory analysis). The last three columns in Table 8.2 give an indication of the relative feasibility (high, moderate, or low) of collecting a certain data type at each of the three scales under consideration.

Due to the large size of areas to be studied at the regional scale (on the order of 500–2000 km^2), and because of the objectives of hazard assessment at this scale, detailed data collection for individual variables is not a cost-effective approach. Data gathered at this scale is limited to the delineation of homogeneous terrain mapping units, and collection of regional seismic data. At the medium scale, nearly all of the data layers given in Table 8.2 can be gathered for areas smaller than 200 km^2, with exception of detailed soil and groundwater information. At the large scale, where the study area is generally smaller than 50 km^2, all of the proposed data layers can be collected.

Methods of analysis

The following subsections systematically present the techniques for landslide hazard zonation for their use in a GIS. An overview of the required input data is given and the various steps using GIS are mentioned briefly. A recommendation is also given regarding the most appropriate working scale.

Landslide distribution analysis

The most straightforward approach to landslide hazard mapping is a landslide inventory map, based on aerial photo interpretation, ground survey, and/or a database of historical occurrences of landslides in an area. The final product gives the spatial distribution of mass movements, represented either at scale or as points. The maps can be used as an elementary form of hazard map, because they display the location of a particular type of slope movement in an area. They provide information only for the period shortly preceding the date the aerial photos were taken or the fieldwork was conducted. They provide no insight into the temporal changes in mass movement distribution. Many landslides that occurred some time before the photographs were taken may have become undetectable.

In most of the methodologies for landslide hazard assessment, a mass movement distribution map is the most important input map, as it shows the distribution of the phenomena that one wants to predict. The input consists of a field-checked photo-interpretation map of landslides for which recent, relatively large-scale, aerial photographs have been used, combined with a table containing landslide parameters, obtained from a checklist. GIS can perform an important task in transferring the digitized photo-interpretation to the topographic basemap projection using a series of control points and camera information.

The GIS procedure followed is:

- digitizing of the mass movement phenomena, each with its own unique label and a six-digit code containing information on the landslide type, subtype, activity, depth, and site vegetation, and on whether the unit is a landslide scarp or body;
- recoding of the landslide map with the parameters for type or subtype into maps displaying only one type of process.

The method is most appropriate at medium or large scales. At the regional scale, the construction of a mass movement distribution map is very time-consuming and too detailed for procedures of general regional zoning. Nevertheless, when possible it is advisable to prepare such a map also for the regional scale, although with less detail.

Two important considerations arise in this method.

- The accuracy of interpretation of mass movement phenomena from aerial photographs depends on the skill of the interpreter, and the interpretation is subjective. Detailed fieldwork is very important.
- GIS in this technique is used only to store the information and to display maps in different forms (e.g. only the scarps, only slides, only active slides). Although the actual analysis is very simple, the use of a GIS is of great advantage in this method. The user can select specific combinations of mass movement parameters and obtain a better insight into the spatial distribution of the various landslide types.

Landslide activity analysis

A refinement of landslide distribution mapping is the construction of landslide activity maps, based on multi-temporal aerial photo-interpretation. To study the effects of the temporal variation of a variable such as land use, landslide activity maps are indispensable.

The code for mass movement activity which is given to each mass movement phenomenon can also be used in combination with mass movement distribution maps from earlier dates to analyze mass movement activity. Depending on the type of terrain which is studied, time intervals of 5–20 years can be selected. This method of interval analysis offers numbers or percentages of reactivated, new, or stabilized landslides. The following GIS procedures are used.

- The digitized map of recent mass movements is used as the basis for the digitizing of maps from earlier dates. This is done in order to make sure that the landslides which were already present at earlier dates are digitized in the same position.
- Calculation of the differences in activity between two different dates, by comparison of the data from the checklists combined with the map data.
- Calculation of all landslides which were initiated or reactivated in the period between the two photo-coverages.

The most appropriate scales are the medium and the large scales, because of the required detail of input maps, as discussed earlier. The main problems with the landslide activity method are that it is very time-consuming, and that it is difficult to prevent inconsistencies between interpretations from the various dates. The information derived from aerial photos from earlier dates cannot be checked in the field, and will result in greater inaccuracies. The method is represented schematically in Figure 8.4, and an example of the use of the methodology is shown for the large-scale area in Plate 4.

Landslide density analysis

Landslide distribution can also be shown in the form of a density map, showing the percentage cover within mapping units. These mapping units may be terrain mapping units (TMUs), geomorphological units, geological units, etc. This method is also used to test the importance of each parameter individually for the occurrence of mass movements. The required input data consist of a mass movement distribution map, and a land-unit map. If the method is used to test the importance of specific parameter classes, the user decides, on the basis of his field experience, which individual parameter maps, or combination of parameter maps will be used. The following GIS procedures are used for mass movement density analysis.

- Calculation of a bit map (presence/absence) for the specific mass movement type for which the analysis is carried out.
- Combination of the selected parameter map with the bit map through map overlay.
- Calculation of the area percentage per parameter class occupied by landslides.

With a small modification, the number of landslides can be calculated instead of the areal density. In that case a bit map is not made, and the mass

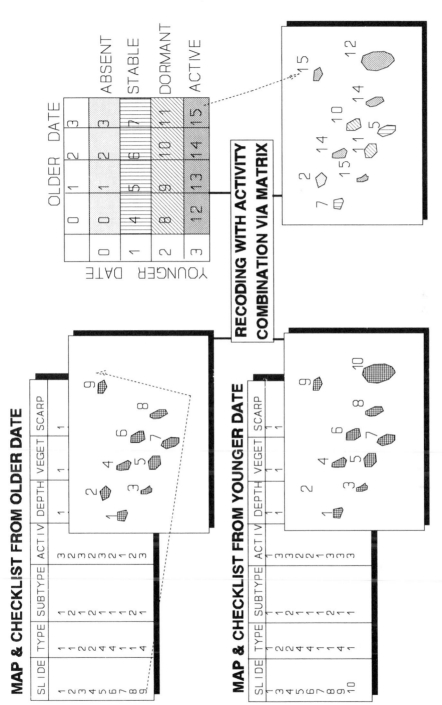

Figure 8.4   Schematic representation of the use of GIS for landslide activity analysis.

movement map itself, in which each polygon has a unique code, is overlaid with the parameter map.

A special form of mass movement density mapping is isopleth mapping (Wright *et al.*, 1974). The method uses a large, moving, counting circle which calculates the landslide density for each circle centre. The result is a contour map of landslide density. The size of the pixels and the size of the filter used define the values in the resulting density map. Except for the creation of a bit map, the procedure for landslide isopleth mapping is rather different.

The method is most appropriate on the medium and large scales for the reasons discussed in the previous section. An example of a series of density maps for the regional-scale area is shown in Figure 8.5.

Geomorphological landslide hazard analysis

In geomorphological methods, mapping of mass movements and their geomorphological setting is the main input factor for hazard determination. The basis for this approach was outlined by Kienholz (1977), who developed a method to produce a combined hazard map based on the mapping of 'silent witnesses (*Stumme Zeugen*)'. In this method, the degree of hazard is evaluated at each site in the terrain. The decision rules are therefore difficult to formulate, as they vary from place to place. Because the hazard analysis is in fact accomplished in the mind of the geomorphologist, geomorphological methods are considered subjective. In this study the terms objective and subjective are used to indicate whether the various steps taken in the determination of the degree of hazard are verifiable and reproducible by other researchers, or whether they depend upon the personal judgment of the researcher. The term subjective is not intended as a disqualification. Subjective analysis may result in a very reliable map when it is executed by an experienced geomorphologist and objective analysis may result in an unreliable map when it is based on an oversimplification of the real situation. Some examples of geomorphological hazard maps can be found in Rupke *et al.* (1988) and Seijmonsbergen *et al.* (1989).

GIS can be used in this type of work as a drawing tool, allowing rapid recoding of units, and correction of units which were coded erroneously. GIS is not used as a tool for the analysis of the important parameters related to the occurrence of mass movements. The method can be applied at regional, medium, or large scales in a relatively short time period. It does not require the digitizing of many different maps. However, the detailed fieldwork requires a considerable amount of time as well. The accuracy of the resulting hazard map will depend completely on the skill and experience of the geomorphologist. Geomorphological maps of the same area made by different geomorphologists may vary considerably, as was tested in this study (van Westen, 1993).

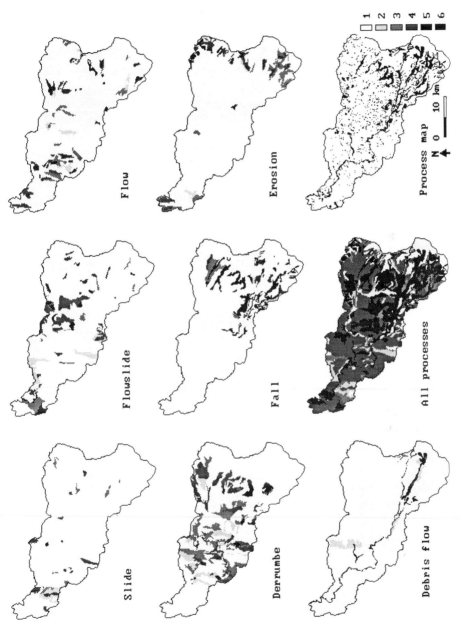

Figure 8.5   Classified density values for several denudational processes calculated per terrain mapping subunit (regional-scale area). Legend: 1 = 0‰, 2 = 0–5‰, 3 = 5–10‰, 4 = 10–50‰, 5 = >50‰, and 6 = all processes.

Qualitative landslide hazard analysis

To overcome the problem of the hidden rules in geomorphological mapping, other qualitative methods have been developed based on qualitative map combination. Stevenson (1977) developed an empirical hazard rating system for an area in Tasmania. On the basis of his expert knowledge on the causal factors of slope instability, he assigned weighting values to different classes in a number of parameter maps. This method of qualitative map combination has become very popular in slope instability zonation. The problem with this method is that the exact weighting of the various parameter maps is often based on insufficient field knowledge of the important factors, which leads to unacceptable generalizations.

The basis for this method of hazard mapping is the field knowledge of the earth scientist who decides which parameters are important for the occurrence of mass movements. Qualitative weighting values are assigned to each class of a parameter map, and each parameter map receives a different weight. Depending on the detail of the study, several input maps can be used, among which the most important are geomorphology, mass movement occurrences, slope angle, geology, land use, and distance to faults, roads, and drainage lines.

The following GIS procedures are used.

- Classification of each parameter map into a number of relevant classes.
- Assignment of weight values to each of the parameter classes (e.g. on a scale of 1–10).
- Assignment of weight values to each of the parameter maps.
- Calculation of weights for each pixel and classification in a few hazard classes.

The method is applicable on all three scales. Each scale has its own requirements as to the required detail of the input maps.

Bivariate statistical landslide hazard analysis

Aiming at a higher degree of objectivity and better reproducibility of the hazard zonation, which is important for legal reasons, statistical techniques have been developed for the assessment of landslide hazard. These quantitative methods have benefited strongly from the availability of computers. Brabb *et al.* (1972) presented a method for quantitative landslide susceptibility analysis at a regional scale, which is based on landslide occurrence, substrate material type, and slope angle. Geological units are grouped according to their landslide density and relative susceptibility values are assigned. Combining these values with a slope map produces final susceptibility classes. The method is easy to use, although it is usually not sufficient to use only the factors of rock type and slope angle.

In this method, overlay of parameter maps and calculation of landslide densities form the core of the analysis. The importance of each parameter, or specific combinations of parameters, can be analyzed individually. Using normalized values (landslide density per parameter class in relation to the landslide density over the whole area), a total hazard map can be made by addition of the weights for individual parameters. The weight values can also be used to design decision rules, which are based on the experience of the earth scientist. It is also possible to combine various parameter maps into a map of homogeneous units, which is then overlaid with the landslide map to give a density per unique combination of input parameters.

GIS is very suitable for use with this method, which involves a large number of map overlays and manipulation of attribute data. This method requires the same input data as the qualitative method discussed in the previous section.

It should be stressed that the selection of parameters has also an important subjective element in this method. The following GIS procedures are used.

- Classification of each parameter map into a number of relevant classes.
- Combination of the selected parameter maps with the landslide map via map overlay.
- Calculation of weighting values based on the cross table data.
- Assignment of weighting values to the various parameter maps, or design of decision rules to be applied to the maps, and classification of the resulting scores in a few hazard classes.

The medium scale is most appropriate for this type of analysis. The method is not detailed enough to apply at the large scale, and at the regional scale the necessary landslide occurrence map is difficult to obtain.

Several specific bivariate statistical methods exist which are based on the same principles, but use different indexes, of which two are briefly described here. The 'information value method' (Yin and Yan, 1988) is a statistical technique that requires a database of parameters collected for different land units. The analysis is based on the presence (1) or absence (0) of landslides at a certain location or within a land unit. It can be used for both alphanumeric and numeric data. The presence or absence of parameters is also calculated. The relative importance for the occurrence of landslides of each parameter is calculated in terms of an information value, which is the log of the landslide density per parameter, as compared to the overall landslide density. In the 'weights of evidence method' (Bonham-Carter *et al.*, 1990), point phenomena (landslides) are regarded along with several terrain factors. These factors are translated into binary input maps. Weights are assigned to the binary maps using Bayes rules for conditional probability. These weights are added to the log of the odds of the prior probability, to give the log of the odds of the posterior probability. The final product of this analysis is a predictor map giving the posterior probability of the occurrence of landslides for each pixel, which is based upon the unique overlap of all binary input pattern maps.

An example of a landslide probability map produced with the weights of evidence method is given in Figure 8.6.

Multivariate statistical landslide hazard analysis

Multivariate statistical analyses of important factors related to landslide occurrence may give the relative contribution of each of these factors to the total hazard within a defined land unit. Carrara *et al.* (1977, 1978) introduced methods for multivariate statistical analysis of mass movement data. Two main approaches of multivariate analysis exist:

1. Statistical analysis of point data obtained from checklists of causal factors associated with individual landslide occurrences (Neuland, 1976; Carrara *et al.*, 1977).
2. Statistical analysis performed on terrain units covering the whole study area. For each of the units, data on a number of geological, geomorphological, hydrological, and morphometrical factors is collected and analyzed using multiple regression or discriminant analysis (Carrara *et al.*, 1978, 1990, 1991; Carrara, 1983, 1988, 1992).

These methods are rather time-consuming, for both data collection and data processing. The analyses are based on the presence or absence of mass movement phenomena within these land units, which may be catchment areas, interpreted geomorphological units or other kinds of terrain units.

Several multivariate methods have been proposed in the literature. Most of these, such as discriminant analysis or multiple regression, require the use of external statistical packages. GIS is used to sample parameters for each land unit. However, with PC-based GIS systems, the large volume of data may become problematic. The method requires a landslide distribution map and a land unit map. A large number of parameters is used, comparable to those mentioned in the previous section. The different classes of a parameter map are considered as individual parameters, resulting in a large matrix. The following GIS procedures are used.

• Determination of the list of factors that will be included in the analysis. As many input maps (such as geology) are of an alpha-numeric type, they must be converted to numerical maps. These maps can be converted to presence/absence values for each land-unit, or presented as percentage cover, or the parameter classes can be ranked according to increasing mass movement density. By overlaying the parameter maps with the land-unit map, a large matrix is created.

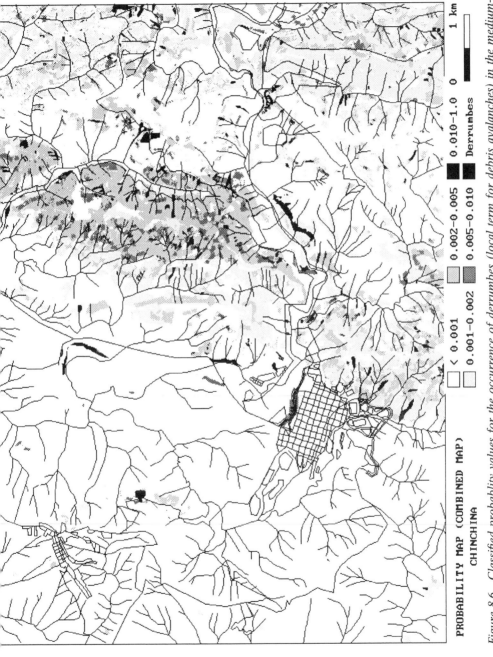

PROBABILITY MAP (COMBINED MAP)
CHINCHINA

☐ < 0.001	☐ 0.002-0.005	■ 0.010-1.0	☐ 0
☐ 0.001-0.002	▨ 0.005-0.010	■ Derrumbes	1 km

Figure 8.6   Classified probability values for the occurrence of derrumbes (local term for debris avalanches) in the medium-scale study area made from a map with unique combinations of variable classes.

- Combination of the land-unit map with the mass movement map via map overlay and dividing the stable and the unstable units into two groups.
- Exportation of the matrix to a statistical package for subsequent analysis.
- Importation of the results per land-unit into the GIS and recoding of the land-units. The frequency distribution of stable and unstable classified units is checked to see whether the two groups are separated correctly.
- Classification of the map into a few hazard classes.

Although these techniques can be applied at different scales, their use becomes quite restricted at the regional scale, where an accurate input map of landslide occurrences may not be available, and where most of the important parameters cannot be collected with satisfactory accuracy. At large scales, different factors will have to be used (such as water-table depth, soil layer sequences and thicknesses). These data are very difficult to obtain even for relatively small areas. Therefore the medium scale is considered most appropriate for this technique.

Deterministic landslide hazard analysis

Despite problems related to collection of sufficient and reliable input data, deterministic models are increasingly used in hazard analysis over larger areas. They are applicable only when the geomorphological and geological conditions are fairly homogeneous over the entire study area and the landslide types are simple. The advantage of these 'white box models' is that they have a physical basis. Their main problem is the high degree of oversimplification. This method is usually applied for translational landslides using the infinite slope model (Brass *et al.*, 1989; Murphy and Vita-Finzi, 1991). The methods generally require the use of groundwater simulation models (Okimura and Kawatani, 1986). Stochastic methods are sometimes used for selection of input parameters (Mulder and van Asch, 1988; Mulder, 1991; Hammond *et al.*, 1992).

Slope stability models require input data on: soil layer thickness; soil strength; depth below the terrain surface of potential sliding surfaces; slope angle; and pore pressure conditions to be expected on the slip surfaces.

The following parameter maps should be available in order to be able to use such models.

- a material map, showing both the distribution at ground surface as in the vertical profile, with accompanying data on soil characteristics;
- a groundwater level map, based on a groundwater model or on field measurements; and
- a detailed slope angle map, derived from a very detailed DTM.

For the application of GIS in deterministic modelling, several approaches can be followed:

- the use of an infinite slope model, which calculates the safety factor for each pixel;
- selection of a number of profiles from the DTM and the other parameter maps which are exported to external slope stability models; and
- sampling of data at predefined grid-points, and exportation of these data to a three-dimensional slope stability model.

The method is applicable only at large scales and over small areas. At regional and medium scale, the detail of the input data, especially concerning groundwater levels, soil-profile, and geotechnical descriptions is insufficient. The variability of the input data can be used to calculate the probability of failure in connection with the return period of triggering events.

The resulting safety factors should never be used as absolute values. They are only indicative and can be used to test different scenarios of slip surfaces and groundwater depths. An example of a landslide probability map is given in Plate 5. This map indicates the probability that the safety factor is lower than 1 when a rainfall event occurs with a return period of 25 years.

Landslide frequency analysis

Most of the methods mentioned so far do not result in real hazard maps as defined by Varnes (1984). Assessing the probability of occurrence at a specific location within a certain time period is possible only when a relationship can be found between the occurrence of landslides and the frequency of triggering factors, such as rainfall or earthquakes. Especially for rainfall-related landslides, various techniques have been developed which determine threshold values of antecedent rainfall (Crozier, 1986; Keefer *et al.*, 1987; Capecchi and Focardi, 1988). Antecedent rainfall is the accumulated amount of precipitation over a specified number of days preceding the day on which a landslide occurred. This permits the calculation of a rainfall threshold value.

The following input data are required:

- daily rainfall records; and
- landslide records (taken from insurance companies, newspapers, or fire/ rescue departments).

The method is most appropriate at medium and large scales. At the regional scale, it may be difficult to correlate known landslides at one location with rainfall records from a different location in the area. The spatial component is usually not taken into account in this analysis, and therefore the use of a GIS is not crucial. GIS can be used to analyze the spatial distribution of

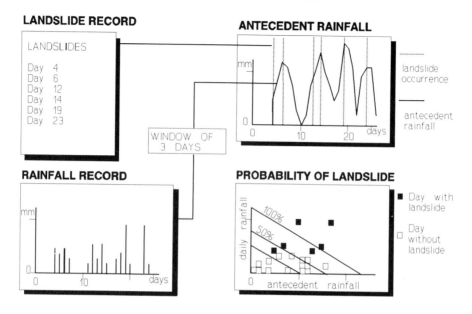

rainfall, however. A schematic representation of the method is given in Figure 8.7.

Discussion and conclusions

Any hazard evaluation involves a large degree of uncertainty. Prediction of natural hazards such as landslides, which are caused by the interaction of factors which are not always fully understood and are sometimes unknown, confronts earth scientists with especially large problems. For large areas and at small, not detailed, scales it is possible to make general predictions: the number of landslides that have occurred in the past within a land unit is a good indication of what can be expected to occur in the near future. It is, however, much more difficult when predictions need to be made in more detail for areas presently free of landslides. In this situation, the earth scientist must rely on models based on the assumption that landslides are more likely to occur in places where a combination of conditions exists which has led to landslides in the past. Most methods presented in the literature and evaluated in this study are based on this principle. This implies knowledge of causal factors, and the ability to represent these on a map, as well as detailed knowledge about past mass movements.

Since hazard maps are used to make predictions over relatively large areas, collection of data for and preparation of these factor maps is a time-consuming operation, and cannot be based solely on factual, measured, field data. During the preparation of these factor maps, the subjective evaluation of field conditions by the earth scientist will play an important role. Since all earth scientists are not equally experienced, these maps will normally contain a considerable degree of uncertainty. It is clear that hazard maps prepared by very experienced geomorphologists will have the highest reliability, with or without the use of GIS. However, solutions must be found to upgrade the reliability of hazard maps in studies where less experienced earth scientists are responsible for the collection of basic data and subsequent analysis. For those cases, it is important to give recommendations as to how the reliability of the end product can be increased, by reducing the uncertainty of the input factors as much as possible. This should be achieved by clear definition of criteria for the interpretation of landslides and their controlling factors, as well as by thorough fieldwork. Instead of making a map by photo-interpretation followed by a field check, input maps for a hazard zonation should be prepared after fieldwork preceded by photo-interpretation.

The use of GIS confronts the earth scientist with the need to provide quantitative values for many uncertainties encountered in the input data, and can serve as an important tool in analyzing the sources of error. It can also help in reducing the errors occurring in the phase of transfer of the photo-information to a topographic map, and in the correct positioning of the various input layers. Apart from the large subjectivity present in the input factors, some of the methods for landslide hazard zonation, evaluated in this study, also contain a considerable subjectivity in the subsequent analytical phase.

GIS offers map overlaying possibilities and calculation facilities far superior to conventional techniques. One of the major contributions of GIS may be the reduction of the subjective element during the analysis phase, allowing the user to concentrate more on reducing errors stemming from the input data. It is especially useful in those situations where the causal factors for mass movements are not fully understood. The user can test hypotheses rapidly, and select the most important combination of factors by trial and error. The result will be optimal when field knowledge is combined with the calculation facilities of GIS. GIS should not be used to throw a large group of variables into a 'black box', to see what comes out, since such an approach is not based on a clear understanding of the causal mechanisms of slope failures. Standard calculation methods are presented, but the user is fully responsible for the selection of relevant input data and the analytical model.

The methods presented in this study cannot be executed at each scale of analysis. Before starting a hazard evaluation study, an earth scientist should be aware of the desired degree of detail of the hazard map, given the requirements of the study. When a degree of detail and a working scale have been defined, the cost-effectiveness of obtaining input data must be considered. This chapter provides recommendations as to which kind of data can be collected at each working scale (regional, medium, and large scale). The availability of data determines the type of analysis that can be executed. Table 8.3 gives a summary of the author's conclusions on the feasibility and

Table 8.3   Summary of the feasibility and usefulness of applying GIS-based techniques for landslide hazard zonation on the three scales under consideration

Method	Regional scale	Medium scale	Large scale	Usefulness of GIS in the analysis
Landslide distribution analysis	2–3	3–3	3–3	Intermediate
Landslide density analysis	2–3	3–2	3–1	Intermediate/high
Landslide activity analysis	1–3	3–3	3–3	Intermediate/high
Landslide isopleth analysis	2–3	3–2	3–1	High
Geomorphological landslide hazard analysis	3–3	3–3	3–3	Very low
Qualitative landslide hazard analysis	3–3	3–2	3–1	Low/intermediate
Landslide susceptibility analysis	1–3	3–3	3–2	High
Information value method	1–1	3–3	3–2	High
Weights of evidence method	1–1	3–3	3–2	High
Multivariate statistical analysis	1–2	3–2	3–2	High
Deterministic landslide hazard analysis	1–1	1–2	2–3	High
Antecedent rainfall analysis	2–2	3–3	3–2	Very low

The first number indicates the feasibility (1 = low: it would take too much time and money to gather sufficient information in relation to the expected output; 2 = moderate: a considerable investment would be needed, which only moderately justifies the output; 3 = good: the necessary input data can be gathered with a reasonable investment related to the expected output). The second number indicates the usefulness (1 = of no use: the method does not result in very useful maps at the particular scale; 2 = of limited use: other techniques would be better, 3 = useful).

usefulness of applying the methods discussed in this chapter for the various scales under consideration, and of the usefulness of GIS. The following recommendations are given.

- For very large areas at the regional scale, the best method is the use of terrain classification based on satellite imagery, followed by qualitative hazard analysis using relative weight values obtained from brief field visits.
- For moderately large areas at the regional scale, it is advisable to use terrain classification based on satellite imagery and interpretation of landslides from aerial photos, followed by a density calculation of landslides per mapping unit.
- At the medium scale, the most useful method consists of the collection of important factors related to mass movement occurrence, followed by re-classification and combination into homogeneous units and calculation of quantitative weight values.
- For geomorphologically homogeneous areas, at the large scale the best method is the application of simple slope stability models.

- For geomorphologically heterogeneous terrain at the large scale, the use of detailed geomorphological mapping is considered the best solution.

GIS will play an increasingly important role in the analysis of landslide hazards. It is an important tool in evaluating the accuracy of the input data. With a good database structure and standardized methods of data gathering, the input maps can be greatly improved during the course of a project by the entry of newly collected data. In this way, GIS will not only serve inexperienced earth scientists in the analysis of unknown causal factors of slope instability within a region, but will also enable experienced professionals to create a detailed database which can be useful for many more engineering geological applications other than landslide hazard assessment alone.

GIS is very important in analysing the complex combination of factors leading to slope instability. It allows the use of models which were previously available, but which could not be used because of the large amounts of time involved in their application. One of the most promising applications of GIS in landslide hazard assessment could be the further development of detailed slope instability models, in combination with groundwater models, applied over relatively small areas. Provided that it is used in combination with detailed field knowledge, GIS will enhance the reliability and objectivity of the hazard maps, which therefore will become increasingly important in the decision-making process.

Acknowledgments

This study was carried out in the framework of a project entitled The Use of Geographic Information Systems for Mountain Hazard Mapping in the Andean Environment, sponsored by UNESCO, EEC and the Government of The Netherlands, in which the following institutes participated: the International Institute for Aerospace Survey and Earth Sciences (ITC), the Colombian Geographical Institute (IGAC) and the French Geological Survey (BRGM). The Universities of Utrecht and Amsterdam, as well as the Colombian National University and the University of Caldas, also participated in the project. The following persons are thanked for their valuable contribution: Dr Niek Rengers, Dr Jan Rupke, Dr Rob Soeters, Dr Theo van Asch, Dr Jean Pierre Asté, Mark Terlien, Henk Kruse, Juan B. Alzate, Jose Luis Naranjo, Monica Dunoyer, Ofelia Tafur, John Freddy Diaz, Carlos Enrique Escobar, Dave Niehaus, Pradeep Mool, Achyuta Koirala, Lucia Innocenti, Joost Molenaar, Edwin Koopmanschap, and Eric Mählman.

References

Alzate, J.B. (Ed.) 1992, *Proceedings, 1er Simposio International sobre Sensores Remotes y Sistemas de Informacion Geografica para el Estudio de Riesgos Naturales*, Bogotá, Colombia.

Baez, E.E., Beltran, L.B. and Gonzalez, G., 1988, *Inventory of mass movements observed in the Colombian road network during 1986*, Technical report ID-UN-018A, Research program on landslides in the Colombian road network, Bogotá: Ministry of Public Works.

Bernknopf, R.L., Campbell, R.H., Brookshire, D.S. and Shapiro, C.D., 1988, A probabilistic approach to landslide hazard mapping in Cincinatti, Ohio, with applications for economic evaluation, *Bulletin of the International Association of Engineering Geologists*, **25**, 39–56.

Bonham-Carter, G.F., Agterberg, F.P., Wright, D.F., 1990, Weights of evidence modelling: a new approach to mapping mineral potential in Agterberg, F.P. and Bonham-Carter, G.F. (Eds) *Paper 8–9*, Ottawa: Geological Survey of Canada, 171–83.

Brabb, E.E., 1984, Innovative approaches to landslide hazard and risk mapping, in *Proceedings 4th International Symposium on Landslides*, Toronto, Canada, Vol. 1, 307–24.

Brabb, E.E., Pampeyan, E.H. and Bonilla, M.G., 1972, *Landslide susceptibility in San Mateo County, California*, US Geological Survey Miscellaneous field Studies Map, MF360, scale 1:62 500.

Brabb, E.E., Guzzetti, F., Mark, R. and Simpson, R.W., 1989, The extent of landsliding in northern New Mexico and similar semi-arid regions, in Sadler, F.M. and Morton, D.M. (Eds) *Landslides in a semi-arid environment*, Publications of the Inland Geological Society, **2**, 163–73.

Brass, A., Wadge, G. and Reading, A.J., 1989, Designing a Geographical Information System for the prediction of landsliding potential in the West Indies, in *Proceedings, Economic Geology and Geotechnics of Active Tectonic Regions*, University College, London.

Burrough, P.A., 1986, *Principles of Geographical Information Systems for Land Resources Assessment*, Oxford: Clarendon Press.

Capecchi, F. and Focardi, P., 1988, Rainfall and landslides: Research into a critical precipitation coefficient in an area of Italy, in *Proceedings, 5th International Symposium on Landslides*, Lausanne, Switzerland, Vol. 2, 1131–6.

Carrara, A., 1983, Multivariate models for landslide hazard evaluation, *Mathematical Geology*, **15**, 403–27.

Carrara, A., 1988, Landslide hazard mapping by statistical methods. A 'black box' approach, in *Workshop on Natural Disasters in European Meditteranean Countries*, Perugia, Italy, 205–24.

Carrara, A., 1992, Landslide hazard assessment in *Proceedings 1er Simposio Internacional sobre Sensores Remotos y Sistemas de Informacion Geografica (SIG) para el estudio de Riesgos Naturales*, Bogotá, Colombia, 329–55.

Carrara, A., Pugliese Carratelli, E. and Merenda, L., 1977, Computer-based data bank and statistical analysis of slope instability phenomena, *Zeitschrift für Geomorphologie N.F.*, **21**, 187–222.

Carrara, A., Catalano, E., Sorriso-Valvo, M., Reali, C. and Osso, I., 1978, Digital terrain analysis for land evaluation, *Geologia Applicata e Idrogeologia*, **13**, 69–127.

Carrara, A., Cardinali, M., Detti, R., Guzzetti, F., Pasqui, V. and Reichenbach, P., 1990, Geographical Information Systems and multivariate models in landslide hazard evaluation in *ALPS 90 Alpine Landslide Practical Seminar. 6th International Conference and Field Workshop on Landslides*, Milano, Italy, 7–28.

Carrara, A., Cardinali, M., Detti, R., Guzzetti, F., Pasqui, V. and Reichenbach, P., 1991, GIS techniques and statistical models in evaluating landslide hazard, *Earth Surface Processes and Landforms*, **16**, 427–45.

Carrara, A., Cardinali, M. and Guzzetti, F., 1992, Uncertainty in assessing landslide hazard and risk, *ITC-Journal*, **2**, 172–83.

Choubey, V.D. and Litoria, P.K., 1990, Terrain classification and land hazard mapping in Kalsi-Chakrata area (Garhwal Hinalaya), India, *ITC-Journal*, **1**, 58–66.

Chowdury, R.N., 1978, *Slope analysis*, Developments in geotechnical engineering 22. Amsterdam: Elsevier.

Crozier, M.J., 1986, *Landslides: causes, consequences and environment*, London: Croom Helm.

Einstein, H.H., 1988, Special lecture: Landslide risk assessment procedure, in *Proceedings, 5th International Symposium on Landslides*, Lausanne, Switzerland, Vol. 2, 1075–90.

Fournier D'Albe, E.M., 1976, Natural disasters, *Bulletin of the International Association of Engineering Geologists*, **14**, 187.

Gardner, T.W., Conners Sasowski, K. and Day, R.L., 1990, Automatic extraction of geomorphometric properties from digitial elevation data, *Zeitschrift für Geomorphologie N.F. Supplement*, **60**, 57–68.

Graham, J., 1984, Methods of stability analysis, in Brunsden D. and Prior, D.B. (Eds) *Slope Instability*, New York: Wiley and Sons, 171–215.

Hammond, C.J., Prellwitz, R.W. and Miller, S.M., 1992, Landslide hazard assessment using Monte Carlo simulation, in *Proceedings 6th International Symposium on Landslides*, Christchurch, New Zealand, Vol. 2, 959–64.

Hansen, A., 1984, Landslide Hazard Analysis, in Brunsden, D. and Prior, D.B. (Eds) *Slope Instability*, New York: Wiley and Sons, 523–602.

Hartlén, J. and Viberg, L., 1988, General report: Evaluation of landslide hazard, in *Proceedings 5th International Symposium on Landslides*, Lausanne, Switzerland, Vol. 2, 1037–57.

Huma, I. and Radulescu, 1978, Automatic production of thematic maps of slope instability, *Bulletin of the International Association of Engineering Geologists*, **17**, 95–9.

International Association of Engineering Geologists (IAEG) 1976, *Engineering geological maps. A guide to their preparation*, Paris: Unesco.

Keefer, D.K., Wilson, R.C., Mark, R.K., Brabb, E.E., Brown III, W.M., Ellen, S.D., Harp, E.L., Wielzorek, G.F., Alger, C.S. and Zatkin, R.S., 1987, Real-time landslide warning during heavy rainfall *Science*, **238**, 921–5.

Kienholz, H., 1977, *Kombinierte Geomorphologische Gefahrenkarte 1:10 000 von Grindelward*, Geographica Bernensia G4, Bern: Geographisches Institut.

Kienholz, H., Mani, P. and Kly, M., 1988, Rigi Nordlene. Beurteilung der Naturgefahren und Waldbauliche Priorittenfestlegung, in *Proceedings INTERPREAVENT 1988*, Graz, Austria, **1**, 161–74.

Kingsbury, P.A., Hastie, W.J. and Harrington, A.J., 1992, Regional landslip hazard assessment using a Geographical Information System, in *Proceedings, 6th International Symposium on Landslides*, Christchurch, New Zealand, Vol. 2, 995–9.

Lopez, H.J. and Zinck, J.A., 1991, GIS-assisted modelling of soil-induced mass movement hazards: A case study of the upper Coello river basin, Tolima, Colombia, *ITC-Journal*, **4**, 202–20.

Mani, P. and Gerber, B., 1992, Geographische Informationssysteme in der Analyse von Naturgefahren, in *Proceedings, INTERPRAEVENT 1992*, Bern, Switzerland, Vol. 3, 97–108.

Mulder, H.F.H.M., 1991, *Assessment of landslide hazard*. Nederlandse Geografische Studies, PhD thesis, University of Utrecht.

Mulder, H.F.H.M. and van Asch, T.W.J., 1988, A stochastical approach to landslide hazard determination in a forested area, in *Proceedings, 5th International Symposium on Landslides*, Lausanne, Switzerland, Vol. 2, 1207–10.

Murphy, W. and Vita-Finzi, C., 1991, Landslides and seismicity: an application of remote sensing, in *Proceedings, 8th Thematic Conference on Geological Remote Sensing*, Denver, Vol. 2, 771–84.

Neuland, H., 1976, A prediction model for landslips, *Catena*, **3**, 215–30.

Newman, E.B., Paradis, A.R. and Brabb, E.E., 1978, *Feasibility and cost of using a computer to prepare landslide susceptibility maps of the San Fransisco Bay Region, California*, US Geological Survey Bulletin 1443.

Niemann, K.O. and Howes, D.E., 1991, Applicability of digital terrain models for slope stability assessment, *ITC-Journal*, **3**, 127–37.

Okimura, T. and Kawatani, T., 1986, Mapping of the potential surface-failure sites on granite mountain slopes, in Gardiner, J. (Ed.) *International Geomorphology*, Part 1, New York: John Wiley & Sons, 121–38.

Pearson, E., Wadge, G. and Wislocki, A.P., 1991, An integrated expert system/GIS approach to modelling and mapping natural hazards, in *Proceedings, European conference on GIS (EGIS)*, **26**, 763–71.

Radbruch-Hall, D.H., Edwards, K. and Batson, R.M., 1979, Experimental engineering geological maps of the conterminous United States prepared using computer techniques, *Bulletin of the International Association of Engineering Geologists*, **19**, 358–63.

Rengers, N., 1992, UNESCO-ITC project on mountain hazard analysis using GIS, in *Proceedings, 1er Simposio Internacional sobre Sensores Remotos y Sistemas de Informacion Geografica (SIG) para el estudio de Riesgos Naturales*, Bogotá, Colombia.

Rengers, N., Soeters, R. and Van Westen, C.J., 1992, Remote sensing and GIS applied to mountain hazard mapping, *Episodes*, **15**, 1, 36–45.

Rupke, J., Cammeraat, E., Seijmonsbergen, A.C. and Van Westen, C.J., 1988, Engineering geomorphology of the Widentobel catchment, Appenzell and Sankt Gallen, Switzerland. A Geomorphological inventory system applied to geotechnical appraisal of slope stability, *Engineering Geology*, **26**, 33–68.

Seijmonsbergen, A.C., Van Westen, C.J., and Rupke, J. (Eds) 1989, *Geomorphological, Geotechnical and Natural Hazard maps of the Hintere Bregenzerwald area (Vorarlberg, Austria)*, Stuttgart: Gebruder Borntraeger.

Soeters, R., Rengers, N. and Van Westen, C.J., 1991, Remote sensing and geographical information systems as applied to mountain hazard analysis and environmental monitoring, in *Proceedings, 8th Thematic Conference on Geologic Remote Sensing*, Denver, Vol. 2, 1389–402.

Stakenborg, J.H.T., 1986, Digitizing alpine morphology. A digital terrain model based on a geomorphological map for computer-assisted applied mapping, *ITC-Journal*, **4**, 299–306.

Stevenson, P.C., 1977, An empirical method for the evaluation of relative landslide risk, *Bulletin of the International Association of Engineering Geologists*, **16**, 69–72.

Van Asch, T.W.J., Van Westen, C.J., Blijenberg, H. and Terlien, M., 1992, Quantitative landslide hazard analyses in volcanic ashes of the Chinchina area, Colombia, in *Proceedings, 1er Simposio Internacional sobre Sensores Remotes y Sistemas de Informacion Geografica para el estudio de Riesgos Naturales*, Bogotá, Colombia, 433–43.

Van Dijke, J.J. and Van Westen, C.J., 1990, Rockfall hazard: a geomorphological application of neighbourhood analysis with ILWIS, *ITC-Journal*, **1**, 40–4.

van Westen, C.J., 1989, ITC-UNESCO project on G.I.S. for Mountain Hazard Analysis, in *Proceedings, First South American Symposium on Landslides*, Paipa, Colombia, 214–24.

van Westen, C.J., 1992a, Scale Related GIS techniques in the analysis of landslide hazard, in *Proceedings, 1er Simposio Internacional sobre Sensores Remotos y Sistemas de Informacion Geografica (SIG) para el estudio de Riesgos Naturales*, Bogotá, Colombia, 484–98.

van Westen, C.J., 1992b, Medium scale hazard analysis using GIS, in *Proceedings, 1er Simposio Internacional sobre Sensores Remotos y Sistemas de Informacion Geografica (SIG) para el estudio de Riesgos Naturales*, Bogotá, Colombia, Vol. 2.

van Westen, C.J., 1993, *Application of Geographic Information Systems to Landslide Hazard Zonation*, ITC-Publication 15, Enschede: ITC.

van Westen, C.J. and Alzate, J.B., 1990, Mountain hazard analysis using a PC-based GIS, in *Proceedings, 6th International Congress of the International Association of Engineering Geologists*, Vol. 1, 265–72.

Varnes, D.J., 1984, *Landslide Hazard Zonation: a review of principles and practice*, Natural Hazards 3, Paris: Commission on Landslides of the IAEG, Unesco.

Wright, R.H., Campbell, R.H. and Nilsen, T.H., 1974, Preparation and use of isopleth maps of landslide deposits. *Geology*, **2**, 483–5.

Yin, K.L. and Yan, T.Z., 1988, Statistical prediction model for slope instability of metamorphosed rocks, in *Proceedings, 5th International Symposium on Landslides*, Lausanne, Switzerland, Vol. 2, 1269–72.

9

Simulation of fire growth in mountain environments

**Maria J. Vasconcelos, José M.C. Pereira and
Bernard P. Zeigler**

Wildland fire in mountainous regions

Forest and shrub ecosystems are the dominant land cover in many mountainous regions where steep topography and poor soils limit the potential for agricultural or urban land use. Wildland fires are a common phenomenon in many types of forests and shrublands and often occur in mountainous terrain. Topographic factors are very influential determinants of wildland fire behaviour, particularly direction and rate of fire spread (Barney *et al.*, 1984). Fireline intensity and flame length are influenced mostly through their dependency on rate of spread. Slope is the most common and direct influencing factor, since a fire burning on an incline has a bent flaming zone, equivalent to that produced by winds (Cheney, 1981; Pyne, 1984). Inclination of the flame front over the unburnt fuel increases heat transmission efficiency and accelerates fire spread. Slopes also encourage spotting by firebrands, carried upslope by convective winds. Topography affects fuel loading and moisture content through altitudinal control of temperature and precipitation, which also determine the length of the fire season.

Position on a slope affects the exposure of fuels and flame fronts to winds and solar radiation, whose fuel drying and fire intensification effects are stronger in the upper third of long slopes (Barney *et al.*, 1984). Ridges and summits are more susceptible to the effects of jetstream winds and to the development of upslope winds on the lee side (Pyne, 1984), and are also more prone to lightning strikes (McRae, 1992). The middle third of a long slope can experience reversals in wind direction up- and down-slope due to differential cooling during night and day, producing erratic, highly unpredictable fire behaviour during the reversal periods. The lower third of a slope receives less wind, less solar radiation and has more lush vegetation, conditions which normally imply less severe fire behaviour (Barney *et al.*, 1984).

The collection of cold air in valleys causes evening inversions of the atmospheric thermal profile which have mixed consequences upon fire behaviour. A stable atmosphere near the surface tends to reduce the rate of spread and intensity of a fire, but it also traps smoke and pyrolysed gases, interfering with fire control operations (Pyne, 1984). In mountainous country, this pooling of colder, denser air in basins and valleys can aggravate burning conditions above the inversion zone. Usually, a thermal belt is formed just above the inversion and a band of strong winds may develop in the same region. This is due to a reduction of thermal and mechanical turbulence at the surface of the mountain, whose roughness is attenuated by the filling of terrain depressions with dense air. Higher windspeeds in this region may unexpectedly increase fire severity (Pyne, 1984).

Slope orientation, or aspect, also affects the amount of exposure of fuels to the drying effects of solar radiation. North and east slopes are more shaded during the day and remain moist and cool for longer periods. Southern and western slopes are more exposed during the warmest part of the day, causing fuels to heat up and dry out, potentially increasing fire severity (Barney *et al.*, 1984).

Irregular terrain leads to microclimates capable of supporting distinct types of fuel, whose loadings and moisture levels may contrast with those of surrounding areas (Pyne, 1984). Narrow valleys and canyons facilitate the formation of convective columns above fires, that act like chimneys. They also facilitate radiant heat transfer across slopes and tend to burn as a unit, with radiant heat or firebrands from one side igniting the other. Natural barriers to fire, such as lakes, streams, swamps, ridges, and rocky outcrops are also strongly related to topography (Barney *et al.*, 1984), and increase the spatial heterogeneity of the fire environment.

Spatial heterogeneity of forest fires

The complex relationships between topography, meteorology and vegetation that are characteristic of mountainous regions produce highly heterogeneous environmental settings for the propagation of forest fires. Most studies analyzing the spatial heterogeneity of forest fires present results concerning the variability of temperature distributions, or fire effects, in burned areas. Heterogeneous temperature fields at the soil surface have been reported by Hobbs, Curral, and Gimmingham (1984) in heath fires in Scotland, by Stott (1986) in dry deciduous dipterocarp forest fires in northern Thailand, and by Hobbs and Atkins (1988) in various shrub and forest communities of western Australia.

The spatial variability of forest fire effects reflects the heterogeneity of fire behaviour, namely of fireline intensity (FLI) and rate of spread (ROS). These are, themselves, influenced by topographic factors, fuel distribution, and weather conditions. Koch and Bell (1980) observed fine-scale spatial variability of fire effects in leaves of *Xanthorroea gracilis*, an undergrowth shrub of Eucalyptus forests in western Australia. Auld (1986), also working in Australia, reports variations in the regeneration of *Acacia suaveolens* throughout burned areas, in a way related to FLI and residence time.

From a standpoint of mountainous environments, the most interesting case study is presented by van Wilgen, le Maitre, and Kruger (1985), who compared predictions of Rothermel's (1972, 1983) fire behaviour model, with fourteen experimental fires in South African fynbos, a Mediterranean-type sclerophyllous shrub community. Fynbos ecosystems exhibit strong environmental heterogeneity, which the authors blame for the relatively poor model performance, especially concerning predictions of FLI. The complex topography and structural heterogeneity of fynbos violate fundamental assumptions of the Rothermel model. Therefore, it is not surprising that there was a better fit between observations and predictions made with the same model in the more homogeneous savanna ecosystems of northern Natal, South Africa (van Wilgen and Willis, 1988).

GIS-based fire behaviour modelling

Awareness of the need to account for the spatial variability of wildfire fire behaviour has led several authors to approach the subject from multiple conceptual and methodological perspectives (Pereira and Vasconcelos, 1990). In the present chapter we review only some of the GIS-based modelling strategies for simulating the spread of fire in heterogeneous environments and discuss in greater detail the application of geographic information systems (GIS) to fire spread modelling.

GIS-based modelling of fire behaviour is different from the spatial modelling approaches reviewed in Pereira and Vasconcelos (1990), in that it often involves the use of real landscapes, with precise geo-referencing of environmental variables, and not the aggregated statistical treatment of actual or simulated spatial heterogeneity. When the spread of specific, real fires is attempted, geometric fire characteristics are not imposed a priori, but result from the mechanics of underlying models. Also, simulation of fire characteristics under alternative vegetation management strategies and weather conditions is made possible by the cartographic data management capabilities of GIS (Vasconcelos and Pereira, 1991).

Salazar and Palmer (1987), Salazar and Power (1988), and Hamilton, Salazar, and Palmer (1989) discussed the usefulness of GIS technology to integrate the large amounts of information required to model fire behaviour in a spatially explicit way. They used a GIS database, containing information on topography, weather, and vegetation types, to calculate rate of spread and fireline intensity for a study area in California. The GIS also served to produce 3D displays of ROS and FLI, and perspective views of bulldozer firebreak production rates, draped over a 3D digital terrain model. These authors did not attempt to simulate the actual spread of fires, ignited at a precisely geo-referenced location, and growing under known meteorological conditions.

McKinsey's (1988) study is very similar, deriving maps of potential FLI from a GIS database containing digital elevation, vegetation type, and vegetation age data. FLI was calculated with two models: one developed at the Boise Interagency Fire Center (BIFC) and Rothermel's (1972) model. This information was used to identify land units of similar expected fire behaviour, and define priority areas for a programme of prescribed burns. However, no attempt was made to model the spread of specific fire events.

Holder *et al.* (1990) addressed the latter problem, by coupling two Canadian fire spread models using raster data and based on cellular diffusion processes, with an analytically oriented GIS, TYDAC Technologies' SPANS. Both models require three sets of inputs: digital elevation data, weather data, and fuel types data. Fuel typing was based on the classification set by the Canadian Forest Fire Behaviour Prediction System (FBP). Three study areas corresponding to different spatial scales were used, to take into account the variation in scale and resolution which fire managers actually have to deal with. Simulation of fire spread was only attempted for the smallest study area, where highest resolution data were available. Maps of the relevant environmental variables contained in SPANS were input to the fire growth models. These calculate fire characteristics, and results are written back to SPANS and displayed graphically. With this approach, fire spread calculations are done within the fire behaviour model, and the GIS is only used to manage the digital cartographic database and to produce graphic outputs. The authors did not clarify the nature of the algorithms used by the fire growth models, other than mentioning that they were based on cellular diffusion processes.

Vasconcelos (1988), Pereira and Vasconcelos (1990), Vasconcelos and Pereira (1991), and Vasconcelos and Guertin (1992) dealt specifically with the coupling between the BEHAVE fire modelling system (Burgan and Rothermel, 1984; Andrews, 1986), which is based on Rothermel's rate-of-spread equation (Rothermel, 1972), and the Map Analysis Package (MAP) GIS (Tomlin, 1986). The FIREMAP system that implements this connection, was used to compare the growth of an actual fire in the White Mountains of central Arizona with model predictions (Vasconcelos, 1988; Vasconcelos and Guertin, 1992), and to analyse the consequences of hypothetical alternative reafforestation projects on fire behaviour (Vasconcelos and Pereira, 1991).

The procedure for dynamic, geo-referenced simulation of fire behaviour involves three separate stages (Vasconcelos, 1988). First, the GIS is used to partition the study area into internally uniform cells, and disaggregate the geographic information into thematic maps corresponding to the different inputs required by the BEHAVE system.

Next, these data are supplied to BEHAVE. Instead of a single vector of spatially-averaged data concerning mean values for topography, weather, and fuels, one such vector is processed for each database cell, in a distributed and precisely geo-referenced manner. Results are then written to corresponding cells and output in cartographic form. Figure 9.1 is an illustration of this process, showing the production of ROS, FLI and direction of maximum spread (DMS) maps.

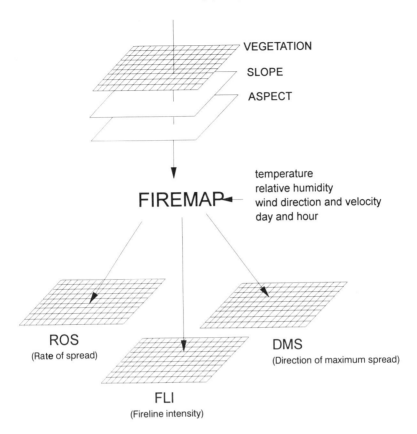

Figure 9.1 FIREMAP generates all weather related overlays and creates the maps with the distributed values of predicted fire characteristics.

The third phase uses the MAP SPREAD algorithm. Generically, GIS SPREAD-type functions are used to evaluate transportation time or cost over a complex surface. These algorithms evaluate phenomena that accumulate with distance by moving outwards in all directions from one or more source points. In the process, a variable such as travel time is calculated and a running total of the calculations is kept. Travel-time data layers define the time it takes to traverse each cell and, if multiple paths are available, the SPREAD function selects the one corresponding to the shortest distance or least-cost path. These algorithms require that the study area be regularly subdivided into relatively small terrain units, to provide the cells over which values are progressively accumulated. Because of this, SPREAD functions are only implemented in raster-based GIS (Aronoff, 1989).

In the present case, fire is the phenomenon being transported or spreading in space, and the travel-time map is obtained by dividing the values in the ROS map by cell size, which is constant for the entire database. This produces a map of the time it takes the flames to traverse each cell. One cell is selected as the point of fire ignition and overlaid with the flame front travel-time map. Subsequently, the SPREAD algorithm processes the data in the way indicated above until the total time chosen for the simulation is accumulated. Results of the simulation include location, shape, and size of burned areas and fire perimeter at user-specified intervals. The system has the ability to accept changes of weather conditions at the beginning of each time interval. Maps of all other fire characteristics calculated by BEHAVE (flame length, heat yield by unit area of the fire front, etc.) may also be produced (Vasconcelos, 1988).

Within each constant weather time interval, all cells are assumed to burn in the same direction, which can only be updated between time intervals. Additionally, there is a need to manually update the fire perimeter at the end of each time interval, so that only those cells burning in the fire front are kept burning and thus influence the next time period (see Vasconcelos and Guertin, 1992 for details).

The application of the described methods in the Ivins Canyon case study (Vasconcelos, 1988; Vasconcelos and Guertin, 1992) showed that, despite the adequacy of GIS as a spatial database management system, there are important limitations to the implementation of dynamic models in a GIS environment. The main problems relate to the lack of flexibility of GIS spatial operators and to the discrete time nature of the simulations. These limitations impose the above-mentioned assumptions of a single rate of spread value associated with each cell and constancy of spread direction during each time interval. Additional limitations are related to the difficulty of using operators applicable only to individual grid cells, and to the lack of flexible rule-based operators. Moreover, GIS are static and do not include procedures for handling time.

The complexity of the fire spread process calls for more powerful models, able to handle both the spatial interactions involved in the spread of fire over an heterogeneous landscape, and the temporal dynamics introduced by varying weather conditions. Green *et al.* (1989) discuss modelling of dynamic landscape processes and consider cellular automata as a promising framework to accomplish this task. Forest fires are mentioned as a typical example of such processes. Good presentations of cellular automata and their potential for modelling landscape dynamics can be found in Couclelis (1985) and Hogeweg (1988). Couclelis (1985) also discusses the work of Zeigler (1976) on distributed discrete event simulation (DEVS), as a more generic and flexible framework that offers an excellent starting point to approach the task of spatial dynamic modelling. Discrete event specifications offer important advantages over cellular automata for modelling spatial dynamic processes in cellular spaces. For a complete discussion and proof see Zeigler (1984).

Vasconcelos and Zeigler (1993) and Vasconcelos *et al.* (in press) present a methodology that links DEVS-Scheme, a knowledge-based discrete-event simulation environment with GIS, and show how it can be used to simulate succession in a landscape subject to recurrent fires. Their methodology can be easily extended to many other spatial dynamic studies with GIS, including the propagation of forest fires in highly heterogeneous regions like mountainous areas.

DEVS-Scheme is an intelligent environment that can associate stand-alone models with particular spatial positions and couple those models in a coherent manner depending on the problem at hand. If needed, there may be a model (an independent processor) associated with each cell of the GIS data base, and DEVS-Scheme can emulate parallel processing with those models in a computationally efficient way. Linking process models of fire spread and surface winds, with vegetation and topographic GIS databases in the framework of DEVS can certainly be a complex, demanding task, but one that, as we hope to illustrate, could represent a qualitative leap in our ability to model the spread of forest fires.

The Discrete Event Systems specification for fire spread modelling in GIS

The methodology we propose, called knowledge-based simulation (Zeigler, 1990), integrates discrete event simulation formalisms and artificial intelligence knowledge-representation schemes, in DEVS-Scheme, with a geographic information system (GIS). DEVS-Scheme is a knowledge-based, object-oriented simulation environment for modelling and design that facilitates construction of families of models in a form easily reusable by retrieval from a model-base. It is engineered in a set of layers so that all of the underlying Lisp-based and object-oriented programming language features are available to the user (Zeigler, 1990).

All models in DEVS-Scheme are hierarchical, modular, object-oriented models. The term modularity means the model is described so that it has recognized input and output ports through which all the interaction with the external world is mediated (Zeigler, 1990). For a set of component models (objects), a coupled model can be created by specifying how the input and output ports will be connected to each other, and to external ports. The new coupled model is itself a modular model and thus can be used as a component in a yet larger, hierarchically higher level model.

Each modular model is an object. Each object has its own variables and procedures to manipulate these variables called methods. Only the methods owned by the object can access and change the values of its variables, and thus its state. Objects can communicate with each other and with higher levels of control, to cause changes in their states, by a process called message passing (Zeigler, 1990). This process uses couplings (input and output ports connected together) as communication channels.

The most basic models from which all others are built by coupling are called atomic models. Atomic models are specified in the dynamic discrete event formalism. Since the DEVS formalism includes closure under coupling (Zeigler, 1984, 1990) all models created by coupling of atomic models (coupled models) are also discrete event models. Atomic models are stand-alone modular objects that contain a set of state variables and parameters, an internal transition function that computes the next state and state transition time when no messages arrive in the input ports, an external transition function that computes the next state and transition time when an external event arrives in an input port, a time advance function, and an output function which generates an output just before an internal transition takes place.

Two state variables are usually present in atomic models: phase and sigma. In the absence of external events the model remains in the current phase for the time given by sigma. When an external event occurs, the external transition function places the system in a new phase and sigma, thus scheduling it for the next internal transition (Zeigler, 1990). The next state is computed on the basis of the present state, the input port and value of the external event, and the time elapsed in the current state.

For the simulation of fire growth in a cellular space, one can envision the placement of an atomic model at each cell location. Thus there may be an array of spatially-referenced atomic models that correspond to distributed processors in the raster map lattice. The coupling of atomic models in space can then be dynamically managed by an extra atomic model, a controller (see below).

The DEVS knowledge representation scheme, the system entity structure, combines decomposition, coupling, and taxonomy, which deals with the admissible variants of a component and their specializations (Zeigler, 1990). The entities of the system entity structure refer to the conceptual components of reality for which models may reside in the model base. In the fire growth case, the models refer to cells in a raster GIS, corresponding to parcels of land.

This knowledge base is a compact representation scheme that can be unfolded to generate the family of all possible models synthesizable from the models in the model base, expressed as composition trees (Zeigler, 1990). For example, the system entity structure for the fire growth model (Figure 9.2) may generate a composition tree with any number of atomic-models depending on the array size and simulation objectives.

DEVS-Scheme includes capabilities of variable structure (Zeigler *et al.*, 1990). This means that a model's initial structure may change as the simulation proceeds, in a way that differs from simulation to simulation, depending on the specific conditions. In the fire growth model, this ability is translated in the possibility of having only models corresponding to active (burning) cells in the model structure. New models are loaded when new cells start burning. Conversely, when a cell burns out and its model is no longer needed, it is removed from the model structure and memory is freed up for the functioning of newly ignited cells.

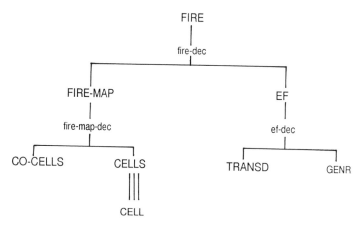

Figure 9.2 The system entity structure for the fire growth model. The extension 'dec' denotes the decomposition of an entity. Three vertical lines indicate a multiple entity, such as CELLS, which can be composed of any number of CELL entities. See also Figure 9.4

The DEVS-GIS model for simulating fire growth in heterogeneous environments

Conceptualization

The first step in the conceptualization of the fire model consists of describing fire behaviour in one homogeneous cell, and establishing the rules governing that behaviour. Then, the interactions of a burning cell with its neighbours are considered.

Once a cell is ignited by contagion, it burns until the fire has crossed the distance corresponding to that cell. Since the rate of fire spread for the considered weather scenario is known (through Rothermel's equation), one can calculate the time it takes to consume the cell, given a direction of spread. Thus, it is possible to generate the time segment, burning time corresponding to the state burning for each cell. Moreover, given an ignition cell, it is possible to compute the time a fire takes to cross the boundaries of several consecutive cells, by generating a time advance function based on the rates of spread of those cells. The crossing of a boundary can then be envisioned as an internal event generated after the cell remained in burning phase from ignition time to ignition time plus burning time.

If there is a wind shift while a cell is burning, the fire rate of spread and direction of maximum spread (DMS) change with it. This means that the burning cell now burns at a different rate, and thus the remainder of the burnable area takes a different time to consume than initially calculated. The time a cell is in burning phase can be updated by computing the proportion of the cell burned before the wind shift based on the elapsed time and previous

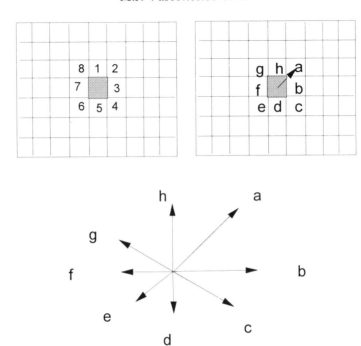

Figure 9.3   Directions in a grid space are represented by considering the immediate neighbourhood of a spreading cell. The cells numbered 1–8 represent the fixed directions to which a cell can spread. The letters represent relative directions and associated rates of spread. Direction a always corresponds to the direction of maximum spread and, if a cell is burning in that direction it is burning at its maximum rate. Conversely, a cell burning in direction e is burning at its slowest rate. The set of relative direction varies from cell to cell and is given by direction of maximum spread in degrees clockwise from uphill.

burning time. The time to burn can than be reset to the time it takes to consume the remainder of the cell at the new rate. This is the proportion left to burn multiplied by the burning time under the new conditions.

The above description for fire in a cell can be encoded in an atomic model CELL. A CELL may have passive, burning, and burned phases with corresponding sigmas of infinite, burning time, infinite. In order to represent fire spread it is necessary to specify how the directions of spread are coded in a grid space and how the varying burning rates in the discretized landscape are linked to produce a coherent total burn. The eight rates of spread (for the directions of the compass, of which one corresponds to the rate of maximum spread) are converted to burning time. Since the cell size is fixed and a given, it is trivial to compute the time a cell takes to get burned in each of the eight directions by dividing the cell's linear size (metres) by the rate of spread in each direction. For directions of burn that correspond to the diagonal of a cell, the time calculated is multiplied by 1·4 to account for a larger linear distance.

The information regarding directions in a grid space is represented as shown in Figure 9.3. The set of numbers refer to the absolute position of the neighbours of any given cell. This scheme facilitates encoding of contagion directions. For example, if the cell is ignited by cell 7 the direction of the burn in this front is directed to the east (to cell 3).

The set of directions represented by letters are relative to DMS. Direction a always represents the direction of maximum spread, so if a cell is burning in direction a (to cell 2, NE in the example) it is burning at the maximum rate and it is a head fire cell. Likewise, if it is burning to cell 5 it is burning at rate *d*. We can envision the letters as representing classes of rates of spread where the letters always represent the spread rate corresponding to a fixed angle from DMS, which varies from cell to cell.

When the time for a cell to burn has elapsed, its neighbours can be ignited, and those that are not already burning will start burning at a rate dependent on the direction of contagion, direction of maximum spread and on the rate at which the source cell was burning. Each cell can set itself to burn at a rate and direction dependent on the information received from the source cell and on its own spread conditions.

The most important assumption made in this first DEVS model concerns the cells that are ignited by source cells burning in the direction of maximum spread. When ignition comes from a source cell burning at its maximum rate, and the direction of contagion matches the direction of maximum spread from the source, then the igniting cell burns at the maximum rate. Additionally, the igniting cell is able to shift the direction of spread to its own direction of maximum spread, becoming a head fire cell.

If a cell is ignited by a neighbour burning in a direction other than DMS, or the direction of contagion does not match the DMS of the source neighbour, it keeps the same direction of spread and it sets the rate of spread to the rate corresponding to that direction. Note that this is not necessarily the same rate class (a, b, c . . .) of the source, and that the selected rate may be the maximum rate in case the direction of contagion matches the igniting cell's DMS.

General model design

Figure 9.4 provides a complete diagram and explanation of the DEVS fire model. This model of fire spread in a cellular space is designed so that, for each cell of the landscape map, there is a corresponding atomic model CELL. Since the model uses the variable structure capabilities of DEVS-Scheme, the number of CELL models at each moment depends on the position of the source CELLs, elapsed simulation time, and fire spread conditions. The set of CELLs present at any time is controlled by another atomic model, CO-CELLS. This manages the couplings for message passing from burning CELLs to neighbouring ignitable CELLs at event times. These models are coupled to form a kernel model of the controlled model type (Zeigler, 1990). The experimental frame (EF) is a coupled model with two components, TRANSD

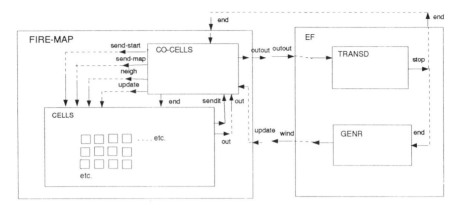

Figure 9.4   Complete box diagram for the fire growth model in DEVS. TRANSD keeps track of the total simulation time and maintains and updates an internal representation of the state of the map that is displayed at event times. Through external and internal couplings of FIRE-MAP and EF, TRANSD receives a message with a list of ignited cells in port 'outout.' To end the simulation, TRANSD sends a message through port 'stop.' When a weather change occurs, GENR sends a message to CO-CELLS through ports 'wind,' 'update,' and 'out.' In turn, CO-CELLS passes it on to all map CELLS with the new values of spread rates and DMS (read from the GIS database) corresponding to updated weather conditions.

The number of CELL models, each represented as small squares in CELLS, varies as the simulation proceeds. CELL models are loaded by replicating the model stored in the model base. Each CELL has output ports 'sendit' and 'outout' through which it acknowledges ignition and reports burnout to CO-CELLS. In addition, CELLS' input ports 'send-start,' 'send-map,' 'neigh,' and 'update' correspond to output ports in CO-CELLS. CO-CELLS sends messages through 'send-start' to initialize new models, and through 'send-map' and 'update' to send information about initial spread conditions and updates, respectively. Through port 'neigh,' CELLs receive information about their spatial position and thus their neighbourhood. This information establishes dynamic couplings for message passing at event times (ignition, burnout, and update).

(transducer) and GENR (generator). EF receives messages from CO-CELLS whenever a CELL of the map undergoes a state transition, and TRANSD updates and displays the map with the new landscape state. The GENR model produces the times for weather updates and sends a message to Co-CELLS, which in turn passes it on to all map CELLS with the new values of spread rates and DMS (read from the GIS database) corresponding to updated weather conditions.

Case study

Site description

An illustration of the concepts introduced above is provided by data from the Ivins Canyon fire of June 1988. Ivins Canyon is located on Spotted Mountain, in the White Mountains range of east-central Arizona, USA. Site elevation ranges from 1700 to 2100 m, and the dominant exposures of the generally steep slopes are to the east, northeast, and southwest.

There are three main vegetation cover types in the Ivins Canyon area: pine stands (*Pinus ponderosa*) with variable crown and understory densities; pine-Douglas fir stands (*Pinus ponderosa-Pseudotsuga mensiezii*); and pinyon-juniper (*Pinus edulis-Juniperus* sp.). The area is under commercial timber management and had been harvested recently. Light, sparse slash was left on the ground. The pine stand understory is mainly of two types: open, with scattered shrubs and a relatively uniform litter layer of pine needles; and brush with Gambel oak (*Quercus gambelii*), in low-quality pine stands, where the crown density of pine is barely above 10 per cent. In the pine-Douglas fir stands, there is a compact layer of short needles and twigs, and new growth is primarily Douglas fir.

GIS database

The GIS database for the Ivins Canyon fire is described in Vasconcelos (1988) and Vasconcelos and Guertin (1992). Briefly, it consists of the following digitized overlays: topography, stream channels, timber type, harvested areas, and fire contours, for an area of nine square miles at a scale of 1:12 000. The digitized vector files were rasterized to a grid of 75 rows by 76 columns, with a cell size of 1 acre (208 ft × 208 ft) on the ground.

Weather data (air temperature, relative humidity, wind speed and wind direction), gathered at the fire camp, were used to generate a set of weather-related map overlays. These were input into the FIREMAP system, together with topography and vegetation data overlays, for calculation of the fire ROS, DMS, and FLI (Vasconcelos and Guertin, 1992).

It should be noted that the weather information was gathered in a standard procedure designed to collect support information for fire fighting. The data were not collected with this kind of study in mind, and some adjustments had to be made to use them in the simulations.

Simulation and results

The simulation of fire spread is done for a four-hour period: 15:00–19:00 on 11 June 1988. The weather data are updated hourly and new sets of ROS, FLI, and DMS are computed hourly. The data available correspond to a portion of a fire that burned from June 10 to 12. Consequently, the initial conditions of the simulation are not optimal because, instead of having a source point in unburnt fuels, there is a large polygon already burned and a fireline whose length is not precisely known.

The simulations start from a source line corresponding to a portion of the perimeter of the previously burned polygon (13:00–15:00) lying in the direction of fire spread, and away from the burned area. This is acceptable, given that the objectives are not to generate an accurate prediction of the growth of this specific fire event, but simply to illustrate the proposed methodology.

Another limitation of this study relates to the manner in which the areas already burned are handled. Since the simulations start with a source line that is considered the fire front, the area lying behind it (to the south) was excluded from the simulation. This results in the blocky appearance of the simulations at the bottom of the burned area. Additionally, there is the limitation of not having fire contours for comparison at the end of each simulation hour. There is only one contour corresponding to the total area burned in the four-hour period.

Figure 9.5 shows the predicted position of the fire perimeter at the end of the first, second, third, and fourth hour. The outline of the actual burned area in the total four hour period is drawn in the background for easier comparison of real and simulated behaviour. The comparison of predicted versus burned areas is shown in Figure 9.6.

The results show a marked overprediction of burned area near the source of the fire. This may be related to the lack of a procedure to extinguish cells. Even though the cells that burn out in reality may correspond to low rates of spread in the model, they are still able to ignite their neighbours. The neighbours may in turn burn well and propagate the fire through a path that actually terminates in a burned cell.

There are possible sources of error that are not related to the dynamics of fire growth, originating in data collection and in the ROS, DMS and FLI predictions by Rothermel's (1972) equation. The accuracy of these calculations does not depend on the dynamic specifications of DEVS, but on the performance of Rothermel's fire behaviour prediction model, on which the DEVS calculations depend. In order to fully appreciate the capabilities of DEVS-GIS for spatial dynamic modelling, it will be necessary to separate the performance of the dynamic implementation in DEVS from the calculation of ROS, DMS, and FLI values.

Discussion and conclusions

The current results are still very limited. However, they illustrate that linking GIS with an intelligent environment, such as DEVS-Scheme, may be a good step towards a more realistic simulation of spatial dynamic processes, such as fire growth.

The discrete event systems specification offers many advantages for simulating fire spread in a discretized space. The problems found in simulations of fire spread within GIS, related to the lack of flexibility of the operators and to the discrete time nature of the simulations, are overcome. The DEVS

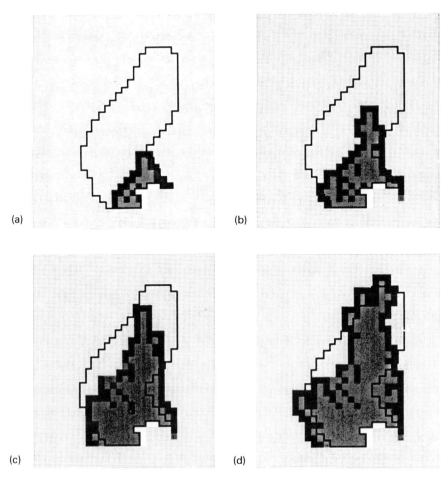

Figure 9.5 Simulation results after one, two, three, and four hours. Black represents burning cells, dark grey represents burned out cells, light grey is the study area, and white is the fire source cells. North is up.

model provides the flexibility for using several rates of spread in the same cell. Under a constant weather scenario, a cell may burn at different rates and in different directions depending on its position relative to the spreading fire. Additionally, each cell may burn in a direction that is different from that of its neighbours. A DEVS-GIS environment will also facilitate linkage with models for other dynamic processes relevant for fire growth modelling, such as models for surface winds over mountainous terrain.

Since the DEVS model uses a continuous time base, there is no need to use discrete time steps. Thus, instead of displaying maps of burned areas at fixed time intervals that show the patch of cells burned during that time step, maps are displayed at event times without loss of information about when

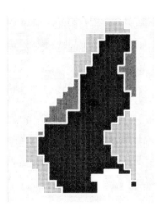

Figure 9.6   Comparison of predicted and actual burned areas. Black represents correct prediction of burned cells. Dark grey represents underpredictions, medium grey represents overpredictions, and light grey is the study area background. White cells are the fire source.

the cells are ignited. Simulations with discrete time bases result in loss of information or unnecessary computation steps (or both). For example, at the end of a time step, there may be cells that are almost totally burned, but are displayed only at the end of the next time step.

In a future version of this model, each cell will calculate the new rates of spread and DMS given the weather conditions (topography and vegetation are parameters in each cell model). This implies that the CELL model will have to incorporate the rate-of-spread equation encoded in its transition functions. This encoding is simple and avoids the computation of large arrays of rates-of-spread values, because each cell can calculate its rates-of-spread values only when activated by an ignition event. The next step in this approach consists of comparing DEVS-GIS simulations with two real, large fires in Portugal, for which there are complete topography, fuels, and weather data sets, ignition points, and burned area maps.

One important aspect illustrated in the DEVS fire model is that, through appropriate representations, one can simulate dynamic, spatially-distributed processes in GIS in a manner equivalent to parallel processing. Moreover, this is achieved by using comparatively small computers (PCs or workstations). The main objections regarding object-oriented approaches to modelling are related to heavy use of computer memory. Although mathematical calculations are reduced to operations dealing with direct responses of entities, the number of operations greatly increases in an object-based model (Sequeira *et al.*, 1991). However, this limitation is significantly reduced in DEVS-Scheme due to its variable structure capabilities, and thus efficient memory usage. Variable structure models are able dynamically to change the models present ·n memory and their couplings, minimizing memory use at any particular time. This is a very important asset when simulating dynamic processes in

GIS because it avoids the need simultaneously to carry large arrays of processing units in memory.

Acknowledgments

This research was supported by the Junta Nacional de Investigaçao Cientifica e Tecnológica, through the funding of research project PMCT/AGR/604/90 and scholarship CIENCIA BD-180 (awarded to the first author). Additional funding was provided by SOPORCEL and CELBI. We also thank the Advanced Resource Technology Laboratory, School of Renewable Natural Resources, University of Arizona, for providing some of the facilities for the early stages of the research.

References

Anderson, H.E., 1984, Calculating fire size and perimeter growth, *Fire Management Notes*, **45**, 25–9.

Andrews, P.L., 1986, BEHAVE: Fire behaviour prediction and fuel modelling system – BURN subsystem. General Technical Report INT-194. Ogden, UT: US Department of Agriculture, Forest Service, Intermountain Research Station.

Aronoff, S., 1989, *Geographic information systems: a management perspective*, Ottawa, Canada: WDL Publications.

Auld, T.A., 1896, Population dynamics of the shrub Acacia suaveolens (Sm.) Willd.: Fire and the transition to seedlings, *Australian Journal of Ecology*, **11**, 373–85.

Barney, R.J., Fahnestock, G.R., Herbolsheimer, W.G., Miller, R.K., Phillips, C.B. and Pierovich, J., 1984, Fire Management, in Wenger, K.F. (Ed.) *Forestry Handbook*, 2nd edn, New York: John Wiley and Sons, 189–251.

Burgan, R.E. and Rothermel, R.C., 1984, BEHAVE: Fire behaviour prediction and fuel modelling system – FUEL subsystem. General Technical Report INT-167. Ogden, UT: US Department of Agriculture, Forest Service, Intermountain Research Station.

Cheney, N.P., 1981, Fire Behaviour, in Gill, A.M., Groves, R.H. and Noble, I.R. (Eds), *Fire and the Australian Biota*, Canberra: Australian Acadamy of Science, 151–75.

Couclelis, H., 1985, Cellular worlds: a framework for modelling micro-macro dynamics, *Environment and Planning A*, **17**, 585–96.

Green, D.G., Reichelt, R.E., van der Laan, J. and Macdonald, B.W., 1989, A generic approach to landscape modelling, in *Proceedings Eight Biennial Conference, Simulation Society of Australia*, Canberra, Australia, 342–7.

Hamilton, M.P., Salazar, L.A. and Palmer, K.E., 1989, Geographic information systems: providing information for wildland fire planning, *Fire Technology*, **25**, 5–23.

Hobbs, R.J. and Atkins, L., 1988, Spatial variability of experimental fires in southwest Western Australia, *Australian Journal of Ecology*, **13**, 295–9.

Hobbs, R.J., Currall, J.E.P. and Gimingham, C.H., 1984, The use of Thermocolor pyrometers in the study of heath fire behaviour, *Journal of Ecology*, **72**, 241–50.

Hogeweg, P., 1988, Cellular automata as a paradigm for ecological modelling, *Applied Mathematics and Computation*, **27**, 81–100.

Holder, G.H., van Wyngaarden, R., Pala, S. and Taylor, D., 1990, Flexible analysis through the integration of a fire growth model using an analytical GIS, in *Proceedings GIS'90 Symposium*, Vancouver, B.C., 152–8.

Kock, J.M. and Bell, D.T., 1980, Leaf scorch in *Xanthorrhoea gracilis* as an index of fire intensity, *Australian Forest Research*, **10**, 113–19.

McKinsey, D., 1988, Priority ranking for prescribed burning in the Cuyamaca Rancho State Park using a geographic information system, in *Proceedings GIS/LIS'88 Symposium*, Vol. 2, 961–70.

McRae, R.H.D., 1992, Prediction of areas prone to lightning ignition, *International Journal of Wildland Fire*, **2**, 123–9.

Pereira, J.M.C. and Vasconcelos, M.J., 1990, Fire propagation modelling in heterogeneous environments and a new spread algorithm for FIREMAP, in *Proceedings, International Conference on Forest Fire Research*, Coimbra, Portugal, B.14.1–15.

Pyne, S.J., 1984, *Introduction to Wildland Fire*, New York: Wiley-Interscience.

Rothermel, R.C., 1972, A mathematical model for predicting fire spread in wildland fuels. General Technical Report INT-115. Ogden, UT: US Department of Agriculture, Forest Service, Intermountain Research Station.

Rothermel, R.C., 1983, How to predict the spread and intensity of forest and range fires. General Technical Report INT-143. Ogden, UT: US Department of Agriculture, Forest Service, Intermountain Research Station.

Salazar, L.A. and Palmer, K.E., 1987, Spatial analysis of fire behaviour for fuel management decision making. Poster paper, in *Proceedings of the GIS'87 Symposium*. ASPRS, Falls Church, VA.

Salazar, L. and Power, J.D., 1988, Three-dimensional representations for fire management planning: a demonstration, in *Proceedings of the GIS/LIS'88 Symposium*, Vol. 2, 948–60.

Sequeira, R.A., Sharpe, P.J.H., Stone, N.D., El-Zik, K.M. and Makela, M.E., 1991, Object-oriented simulation: plant growth and discrete organ to organ interactions, *Ecological Modelling*, **58**, 55–89.

Stott, P., 1986, The spatial pattern of dry season fires in the savanna forests of Thailand, *Journal of Biogeography*, **13**, 345–58.

Tomlin, C.D., 1986, The IBM personal computer version of the Map Analysis Package. GSD/IBM AcIS Project, Report No. LCGSA-85-16 Laboratory for Computer Graphics and Spatial Analysis. Graduate School of Design, Harvard University.

van Wilgen, B.W., le Maitre, D.C. and Kruger, F.J., 1985, Fire behaviour in South African fynbos (macchia) vegetation and predictions from Rothermel's fire model, *Journal of Applied Ecology*, **22**, 207–16.

van Wilgen, B.W. and Wills, A.J., 1988, Fire behaviour prediction in savanna vegetation, *South African Journal of Wildlife Research*, **18**, 41–6.

Vasconcelos, M.J., 1988, Simulation of fire behaviour with a geographic information system. Unpublished MSc thesis, School of Renewable Natural Resources, University of Arizona. Tucson, Arizona.

Vasconcelos, M.J. and Guertin, D.P., 1992, FIREMAP-simulation of fire growth with a geographic information system, *International Journal of Wildland Fire*, **2**, 87–96.

Vasconcelos, M.J. and Pereira, J.M., 1991, Spatial dynamic fire behaviour simulation as an aid to forest planning and management, in Nodvin, S.C. and Waldrop, T.A. (Eds) *Proceedings of the Symposium Fire and Environment*, 421–6. General Technical Report SE-69. Asheville, NC: USDA Forest Service, Southeastern Forest Experiment Station.

Vasconcelos, M.J. Perestrello and Zeigler, B.P., 1993, Simulation of forest landscape response to fire disturbances, *Ecological Modelling*, **65**, 177–98.

Vasconcelos, M.J. Perestrello, Zeigler, B.P. and Graham, L.A., 1993, Modelling spatial dynamic ecological processes under the discrete event systems paradigm. *Landscape Ecology*, **8**(4), 273–86.

Zeigler, B.P., 1976, *Theory of modeling and simulation*, New York: J. Wiley & Sons.
Zeigler, B.P., 1984, *Multifaceted modelling and discrete event simulation*, London: Academic Press.
Zeigler, B.P., 1990, *Object-oriented simulation with hierarchical, modular models*, Boston: Academic Press.
Zeigler, B.P., Kim T.G. and Chilgee, L., 1990, Variable structure modelling methodology: an adaptive computer architecture example, *Transactions Society for Computer Simulation*, **7**, 291–318.

PART III

Research and Resource Management
In and Around Protected Areas

10

Form and pattern in the alpine environment: an integrated approach to spatial analysis and modelling in Glacier National Park, USA

Stephen J. Walsh, David R. Butler, Daniel G. Brown and Ling Bian

Introduction

Our research into the spatial and biophysical character of mountain environments has focused on Glacier National Park, Montana, USA. The Park is characterized by pronounced topographic variability resulting in strong biophysical gradients that shape the corresponding vegetative landscape. The glacial history of the Park, its position astride the Continental Divide, its past geologic setting and present geomorphic processes, its spatial and biophysical complexity, and its relative inaccessibility have necessitated the development of an analytical framework which integrates satellite remotely-sensed data with biophysical coverages contained within a geographic information system (GIS) for spatial analysis and modelling.

Out research within the Park has generally been organized around four primary initiatives:

1. development of a GIS to support spatial and biophysical studies of processes and feature distributions;
2. implementation of resource evaluation studies through the manipulation of the derived GIS database, for the evaluation of natural hazards and snow-avalanche paths; assessment of lake turbidity levels and relationship to basin morphometry; identification of deltaic wetlands; and evaluation of forest fire potential;
3. examination of scale dependencies of plant biomass and topography and spatial analyses of derived patterns of the distributions of specific phenomena; and
4. cartographic and quantitative modelling of disturbance factors and topoclimatic variables affecting the biophysical landscape at the alpine treeline ecotone.

The basic objectives of this chapter are to describe the generation of a centralized GIS database that was designed to accommodate a range of research initiatives within a mountain environment; summarize how the GIS was used to characterize the biophysical landscape and assess derived spatial patterns, through a description of various research initiatives; and discuss how cartographic and empirical models were used to understand the locational and compositional variability of disturbance regimes and the alpine treeline ecotone occurring within the Park. Field-based aspects of the research, employed to validate remote-sensing measures and GIS representations and to establish functional relationships between biophysical variables, are also described. The specific areas of research are discussed with special reference to the role of GIS as the analytical framework for representing, characterizing, and modelling selected elements of the mountain environment.

Study area

Glacier National Park is a UNESCO-designated Biosphere Reserve of 405 000 ha, with peaks to an altitude of 3100 m, in northwestern Montana, USA (Figure 10.1). It was chosen as the study area for our research because of its ecotonal position at the eastern edge of the climatic region dominated by Pacific maritime air in the northwestern United States, as well as for its ecotonal transition from mountains to the Great Plains. The Park also has been selected by the National Park Service to lead the park-oriented scientific investigation of responses to global climate change. In addition, each of the authors had prior field experience within the Park, and currently has ongoing research activities within the study area.

Two mountain ranges, both with glaciers at the present time, occur within the Park: the Livingston Range and the Lewis Range. The Continental Divide follows the Livingston Range in the north and then shifts to the more easterly Lewis Range. The Park area was heavily glaciated during the Pleistocene, but a number of mountain crest and upland surfaces in the Lewis Range escaped glaciation. These gently sloping uplands are largely at or above treeline, with a variety of exposures, and are or have been subjected to periglacial mass-wasting and patterned ground formation in the recent past (Butler and Malanson, 1989). Areas above treeline in the Livingston Range tend to be steeper and more geomorphically active, with little colluvial cover. The periglacial environment in the Park is currently encountered above approximately 1950–2000 m, with current processes producing patterned ground above *ca.* 2170 m, and relict features occurring in the roughly 200 m of transition between 1950 and 2170 m (Butler and Malanson, 1989).

As a result of its location, there are two distinct climatic regions within the Park. West of the Continental Divide, the climate is generally milder and more moist, akin to a Pacific maritime climatic pattern. East of the Divide, the climate is much harsher, with more severe low temperatures, stronger

Figure 10.1   Study area location, Glacier National Park, Montana, USA.

winds, and drier conditions. Snow avalanches include both wet-snow and dry-powder types, but historical data suggest that the former is much more common, especially on western slopes. Major transportation links (US Highway 2 and the Burlington Northern Railroad tracks) pass through avalanche-prone terrain in a narrow valley along the southern boundary of the Park. The peak snow-avalanche month is February, with secondary maxima in January and March (Butler, 1986a). Meteorological triggering mechanisms include:

1. heavy snow;
2. heavy snow followed by a rapid rise in temperature to above freezing;
3. a rise in air temperature to above freezing without precipitation; and
4. rain in association with above-freezing temperatures (Butler, 1986a, 1988).

The high moisture content of the wet-snow avalanches on the western slopes, in particular, imparts a relatively high fluidity to the avalanches, allowing densely packed wet snow to travel easily in path runout zones, causing severe impacts (Butler and Malanson, 1985a). This is especially true where such avalanches are confined within narrow but deeply incised avalanche tracks (Butler and Malanson, 1992; Butler *et al.*, 1992).

Tree species comprising the upper treeline ecotone reflect climatic differences: west of the Continental Divide, upper treeline is comprised primarily of subalpine fir (*Abies lasiocarpa*), Engelmann spruce (*Picea engelmannii*), and subalpine larch (*Larix lyallii*); whereas east of the Divide, treeline is dominated by subalpine fir, whitebark pine (*Pinus albicaulis*), and limber pine (*Pinus flexilis*). Because of the greater amounts of moisture west of the Divide, erosion from running water and past and present glaciers has produced particularly rugged topographic settings. Arno and Hammerly (1984) describe this environment as a combination of extreme topographic and climatic stress. East of the Divide, the drier climate precluded glaciation during the late Pleistocene at some locations, leaving broad, gently sloping uplands at and above treeline (Butler and Malanson, 1989). Upper treeline at these sites is characterized by less stressful growing conditions than those on steep, soil-poor sites west of the Continental Divide. Consequently, treeline at these eastern sites should be able to respond more rapidly to climatic changes brought about by changes in levels of atmospheric carbon dioxide.

Approximately two-thirds of the Park is forested. The subalpine forest grades into the alpine environment, characterized by the diminished vegetation density representative of tundra vegetation and bare rock, snow, and ice surfaces. Plate 6 is a Landsat Thematic Mapper composite image of channels 2, 3, and 4 that shows this grading of vegetation. On the eastern side of the Park, two treelines are evident: an upper temperature-related treeline, and a drought-related treeline at low elevations (Walsh *et al.*, 1989). The spatial extent of this vegetation grading, clearly evident at alpine treeline, is related to elevation, topographic orientation, topoclimatic controls, and disturbances. The eastern two-thirds of the Park are dominated by unvegetated and more sparsely vegetated surfaces. These are interrupted by valleys and hanging-valleys of subalpine forest, brush, and herbaceous cover resulting primarily from the position of the Continental Divide and local orientations of the topography and landforms carved by glaciers (Figure 10.2). Snow-avalanche paths are striking features that extend from the alpine to the subalpine environment throughout the Park (Butler and Walsh, 1990; Butler *et al.*, 1991c). Where vegetation disturbances exist, such as snow-avalanche paths and debris flows, the boundary between vegetated and unvegetated surfaces is distinct, and the transition zone from subalpine forest to alpine conditions is spatially irregular, depending on the pattern and type of disturbance.

Since topographic orientation exerts a substantial effect on vegetation, terrain conditions further serve to modulate the pattern and topology of alpine treeline. The relationship between scale and process is particularly important. Extreme slopes exhibit a narrower transition zone from subalpine forest to tundra vegetation, while gentler slopes show a more gradual and spatially broader progression to the tundra environment. Steeper slopes also show a less organized sequential transition from subalpine forest to krummholz to tundra to bare surfaces. A high degree of variability is apparent in their sequence of occurrence and in their areal extent. On the west side of the Park, a northwest-southeast trending treeline is evident in Plate 6. This is a

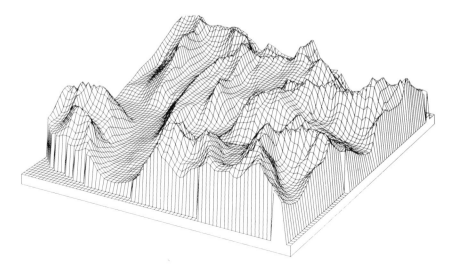

Figure 10.2   A 1:24 000 base-scale digital elevation model (90° direction of view with a 25° angle of tilt) of a portion of the Park (Logan Pass, Montana Quadrangle) showing pronounced topographic variation and evidence of Pleistocene glaciation.

consequence of the topographic variation and corresponding changes in life forms that extend from the closed-canopy forest, to tundra and non-vegetated surfaces.

Treeline in the Park has responded to Holocene environmental changes, but well-dated chronologies are lacking. Elias (1988) used insect fossils to show that treeline at the southern edge of the Park was at least 450 m lower than present at about 11 400 B.P. (before present). This time marked the onset of Pleistocene deglaciation in the area (Carrara, 1987). Although glacial evidence for Holocene climatic deterioration is largely restricted to the Little Ice Age (roughly 1600–1800 A.D. in this area: Carrara, 1987; Butler, 1989a) in the Park, Butler and Malanson (1989) showed that a mid-Holocene cold-climate episode depressed treeline and induced widespread solifluction at relatively low elevations. Treeline was also lowered, but to unidentified levels, during the Little Ice Age. Since the late 1800s, the glaciers of the Park have retreated markedly, and forest advance onto recently deglaciated terrain has occurred. This retreat from Little Ice Age maxima has reduced the hazard from ice avalanches over cliff edges, but increased the danger of catastrophic lake outbursts from moraine-dammed lakes (Butler, 1989a).

Development of a GIS for landscape characterization

The alpine environment of the Park was examined through the development of a GIS in which satellite data sets were processed and integrated with data from collateral sources combined within a GIS framework. The general steps in the development of the GIS included:

1. pre-processing of the satellite data to remove geometric and radiometric distortions in the data;
2. processing of Landsat Multispectral (MSS) and Thematic Mapper (TM) and SPOT Multispectral and Panchromatic digital data to alleviate topographic-induced biases in the spectral responses;
3. enhancement of the satellite digital data – through ratio, vegetation indices, principal components analysis, and spatial filters – to discriminate between vegetation and terrain units;
4. classification of the satellite data to categorize spectral responses for land cover differentiation in the alpine environment;
5. transformation of the remotely-sensed land cover information, processed through ERDAS software, from raster to vector formats for entry into the GIS, processed through ARC/INFO software;
6. analysis of the biophysical landscape by appraising the alpine environment through pattern, process, and scale perspectives contained within the GIS database; and
7. modelling of the pattern and location of alpine treeline and disturbance factors through statistical techniques using data sets and relationships maintained within the derived GIS database.

Table 10.1 lists the variables that were assembled to support the research initiatives, achieved through the processing steps cited above.

The attribute portion of the database was organized in the relational model, whereas locational information was organized within the network model of ARC/INFO. All image processing was accomplished through the ERDAS image-processing software. The Live-Link between ERDAS and ARC/INFO was used to directly overlay satellite data formatted in the raster environment with GIS coverages formatted in the vector environment. Relational join-items were established to link attribute tables within ARC/INFO to support customized database queries. Analytical operations, such as spatial buffers; factor weightings; suitability analyses; data normalizations; and the generation of statistical surfaces, were used to develop additional attribute information and spatial coverages from the initial set of database variables. The spatial operations yielded important information that was permanently added to the database, either as unique coverages with accompanying INFO attributes or as additional fields of information within the attribute database. Data collected in the field were also automated and placed within the database as stand-alone coverages and attribute files. Interfaces between the database and statistical, spatial analysis, and spatial modelling packages were developed to perform various tests and analyses with derived spatial products. For example, displays of regression residuals, measures of spatial autocorrelation, and evaluations of landscape spatial structure through contagion, dominance, and fractal dimensions, were entered into the database for use as stand-alone indicators of spatial pattern and landscape structure and for use in subsequent quantitative analyses. The derived database was used to support the analyses described below.

Table 10.1 Biophysical database developed for the Integrated GIS (USGS = US Geological Survey; DEMs = Digital Elevation Models)

Source	Variable	Scale
EOSAT Corporation	Landcover	30 × 30 metres
SPOT Corporation	Landcover	20 × 20 metres
SPOT Corporation	Landcover	10 × 10 metres
Aerial Photography	Debris Flows	1:34 000 & 1:58 000
EOSAT Corporation	Lineaments	30 × 30 metres
EOSAT and SPOT Corp.	Plant Biomass	30 & 20 metres
USGS DEMs	Elevation	1:24 000 & 1:250 000
USGS DEMs	Slope aspect	1:24 000 & 1:250 000
USGS DEMs	Slope angle	1:24 000 & 1:250 000
USGS	Surficial Geology	1:100 000
USGS DEMs	Curvature	1:24 000 & 1:250 000
USGS	Lithology	1:125 000
USGS DEMs	Talus Slopes	30 × 30 metres
USGS	Hydrography	1:24 000
USGS	Roads & Trails	1:24 000
National Park Serv.	Recent Fires	1:100 000
EOSAT MSS	Snowfields	79 × 79 metres
EOSAT/TM and SPOT	Solar Potential	30 × 30 metres
EOSAT/TM and SPOT	Soil Moist. Potential	30 × 30 metres
Aerial Photography	Snow Avalanche Path	1:34 000 & 1:58 000
Field Sampling	Soil and Plant Types	Local measurements

Resource evaluation of the biophysical landscape

Snow avalanches and other hazards

Snow avalanches represent a major climogeomorphic hazard in Glacier National Park. We have documented numerous cases of property damage, disruption to transportation links, personal injury, and death (Butler, 1986a, 1986b, 1987, 1988; Butler and Malanson, 1985a). Snow avalanches that travel significant distances beyond what is normally expected, and that descend to low elevations where transportation corridors are concentrated, are of special interest because of their hazardous nature. We have examined the topographic conditions most conducive to such avalanches (Butler, 1980; Butler et al., 1991c; Walsh and Butler, 1991; Butler and Malanson, 1992; Butler et al., 1992; Gao and Butler, 1992), and described the hazards associated with snow-avalanche damming of streams along transportation corridors (Butler, 1989c). Much of this work was carried out in association with field verification efforts in support of the development of the GIS database (see especially Butler and Walsh, 1990; Butler et al., 1991c; and Butler et al., 1991d).

In our efforts to determine why snow-avalanche paths occur where they do, we have found it necessary to examine the overall geomorphic effects of snow avalanches in the Park. Although avalanches are strongly associated with topographic depressions such as couloirs (Butler and Walsh, 1990;

Butler *et al.*, 1992; Walsh *et al.*, 1990a), the question arose as to the actual geomorphic efficiency of snow avalanches vis-a-vis stream erosion, past glacial scouring, and periglacial hydrofracturing. Work over a decade on several avalanche paths revealed that very little geomorphic excavation is currently produced by snow avalanches (Butler, 1985, 1989b; Butler and Malanson, 1990), in spite of tree-ring deformational sequences and vegetational patterns in the lateral and longitudinal margins of avalanche paths suggestive of avalanche frequencies on the order of (at minimum) every 3–5 years (Butler, 1985; Butler and Malanson, 1985a, 1985b; Butler *et al.*, 1987).

The vegetational patterns of snow-avalanche paths contrast strikingly with the surrounding mature coniferous forest characteristic of the subalpine environment. The avalanche paths appear as light-green swaths of flexible shrubs and herbaceous plants cutting vertically through the surrounding forest; the relative proportion of shrub and herbaceous cover is a function of avalanche frequency and localized site conditions (Malanson and Butler, 1984a, 1984b, 1985, 1986). These swaths through the forest are easily visible on aerial photographs as well as satellite images. One of the first major steps in our GIS database development was the mapping of snow-avalanche paths from aerial photographs (Butler, 1979; Gao and Butler, 1992), Landsat TM digital data (Walsh *et al.*, 1989; Walsh and Butler, 1989; Walsh *et al.*, 1990a; Butler and Walsh, 1990), and SPOT multispectral and panchromatic digital data (Butler *et al.*, 1992). Land cover classifications derived from the satellite data sets were transformed and then transferred to the GIS database for incorporating the vegetational differences between avalanche paths and coniferous forest. A broad set of path attributes were developed through a combination of the interpretation of aerial photographs and topographic maps, and field observations. The attributes that were generated for each of the 121 snow-avalanche paths used in the analyses and the 34 paths that were used to validate model building and other spatial/biophysical analyses were as follows: length; width; length/width ratio; starting and ending elevation; slope aspect and angle; elevation of ridge above path location and valley below path location; difference between elevation of ridge and valley above and below path location; lithology; distance of path to nearest dike, sill, fault and road; land cover type in starting zone; UTM easting of path head; UTM northing of path head; UTM easting of ridge above path; UTM northing of ridge above path; source and runout area; track length. The source, track, and runout zones of the snow-avalanche paths were interpreted from aerial photography, transferred to a set of topographic maps, and digitized into the GIS as a unique coverage, with a set of accompanying morphometric descriptors of each zone annotated within the INFO portion of the ARC/INFO GIS database (Walsh *et al.*, 1990a).

Other mass movement hazards in the Park have been mapped to complement the work on snow-avalanche paths (Butler, 1979; Oelfke and Butler, 1985a; Butler *et al.*, 1986; Butler, 1989a; Butler, 1990), especially along the eastern margins of the Park where the Lewis Overthrust Fault occurs above

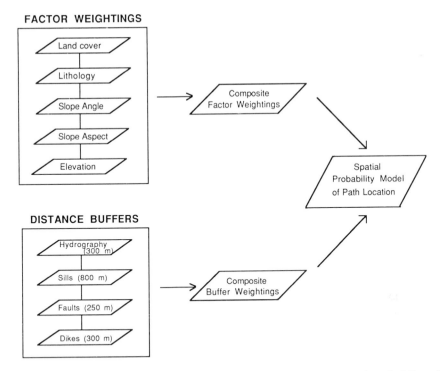

Figure 10.3 Analytical strategy for determining the spatial/biophysical probability of snow-avalanche path location in the Park.

weak Cretaceous mudstones and shales. Dating of several landslide deposits using dendrogeomorphic and lichenometric techniques (Oelfke and Butler, 1985b; Butler *et al.*, 1986; Butler *et al.*, 1991a) has provided a clearer explanation for interpretation of the spectral signals of landslide deposits in the remote-sensing component of the research programme. Current work is examining the landslides along the Lewis Overthrust in relation to the stability of landslide dams (Butler and Malanson, 1993b), and also the relationship of landslides to geologic and topographic controls, following the same general pattern as in our cartographic modelling of snow-avalanche paths (Walsh *et al.*, 1990a).

In the cartographic modelling research, the probability of snow-avalanche path location was modelled through the integration of lithologic, structural, hydrographic, topographic, and land cover variables. Distance buffers were also generated and the attribute database incremented to reflect spatial relationships of paths to hypothesized controlling factors. Factor composites were intersected to determine the spatial probability of path location. All path frequencies within variable classes were normalized through the GIS by the area of class occurrence relative to the total area of the study area, and then added to the morphometric information on each path contained within the INFO tables of the GIS. Figure 10.3 presents a general overview of the processing flow followed in this research. Composite weightings of the biophysical variables influencing path location and weightings of distance measures from selected target features were generated for 121 snow-avalanche paths in order to analyse location probabilities. Plate 7 shows a three-dimensional composite graphic of snow-avalanche paths (with source, track, and runout zones delineated), vegetation productivity (Normalized Vegetation Difference Index: NDVI) derived from Landsat TM data, and a 1:24 000 base-scale digital elevation model (DEM). 'Data organized in the GIS provided an effective approach for model development, evaluation, and variable re-weighting for performance enhancement of the model. The capability to perform spatial proximity measures of paths to selected morphological factors contained within the GIS and to add derived attribute data to INFO tables provided essential information for model building not effectively generated through other approaches' (Walsh *et al.*, 1990a: 615).

Walsh and Butler (1991) reported that a two-way analytical approach is useful for biophysical analyses. In such analyses, the GIS is used to explore biophysical relationships through cartographic models, multiple coverage overlay analyses presented as three-dimensional drapes onto topographic surfaces, and the statistical analysis of biophysical descriptors contained within the GIS that are displayed as statistical coefficients and as GIS graphical surfaces. The ability to maintain a spatial-graphical perspective of statistical measures that are customarily non-spatial is effective for biophysical analyses. Spatial images of regression residuals, measures of spatial autocorrelation, and probability surfaces are useful analytical techniques to explore variable relationships. Also, the capability to validate graphic renditions of variable patterns and relationships, for example through confidence intervals, strengthens the function of the GIS for spatial-biophysical studies.

In reviewing our hazards research over almost two decades, the changing technology that has emerged in GIS in this time period immediately becomes apparent. When Butler (1979) first mapped snow-avalanche paths in Glacier National Park, he did so on acetate overlays with no thought of being able to enter or overlay additional data layers relating to avalanche path location. The emergence of GIS technologies in the 1980s, and continuing to the present, has allowed much more thorough examination of spatial factors that may influence where natural hazards occur or are likely to occur in the future. GIS technology has also tremendously accelerated the speed with which additional hazards can be examined, and allows comparison of spatial distributions of different hazards, their spatial controls, and their locations of likely future occurrence. Given that mountain environments, with their steep slopes, large amounts of unconsolidated debris, seismic potential and frequently moist conditions, are some of the most hazard-prone landscapes on earth, it is not surprising to see that GIS has outstanding capabilities for assisting in our understanding of landscape hazards. What may be surprising is that more has not been done.

Evaluation of wetlands and lake turbidity levels

GIS techniques were employed to study deltaic wetlands associated with finger lakes (Butler *et al.*, 1991b) and the indirect effects of forest fires (Butler *et al.*, 1991d) in the Park. Wetland conditions were explored using aerial photography and satellite digital data that were classified into land cover types and related to vegetational and hydrologic conditions through fieldwork. Compositional differences were discerned between vegetation types characterizing alluvial fans and deltaic wetlands. Statistical clusters generated through the classification of the Landsat TM digital data were related to the chlorophyll content of the defined land cover types and water turbidity levels as an approach for making inventories of deltaic wetlands and for discriminating them from alluvial fans.

Brown and Walsh (1991) modelled lake turbidity levels in alpine lakes by regressing non-synchronous in-situ water quality data with Landsat TM spectral channels. Brown and Walsh (1992) used the morphometry of alpine lake basins to explain lake turbidity estimates derived from satellite-based multiple regression models. Topographic maps and DEMs were used to generate a set of morphometric variables for the analyses. Correlation, clustering, and Principal Components Analysis were used to identify the sets of morphometric variables important in explaining lake turbidity estimates from the satellite data. The GIS was used to consolidate the remote-sensing and topographic information and to establish pathways to statistical packages for subsequent analyses. This research indicated that lake water quality, as measured by satellite multispectral data, can provide information regarding the biophysical processes affecting erosion and sediment yield. Changes in lake turbidity through time, which was shown to be dominated by the extent of glacial ice within a drainage basin, can be measured remotely and analyzed within the GIS framework as an indicator of seasonal, annual, and decadal changes in the impact of glacial ice on lake water quality.

Other research in the riparian habitats of the Park (Butler, 1991; Malanson and Butler, 1990, 1991; Butler and Malanson, 1993a, 1993b) is providing a clearer understanding of the changing spectral signatures associated with plant succession along river corridors, and with the re-establishment of beaver (*Castor canadensis*) in their native range in valley bottoms throughout the Park.

Fire potential modelling and apparent map errors

In the study of forest fire potential, post-fire vegetation succession was examined using satellite and GIS spatial coverages of snow-avalanche paths and landslides to illustrate areas of high, medium, and low fire hazard potential (Butler *et al.*, 1991d). Digital elevation models and satellite digital data were processed and integrated to create elevation, slope angle, slope aspect, and land cover overlays. Using the GIS, fire potential was evaluated through the integration of terrain conditions and the interruption of fuel potential as a consequence of disturbances, snow-avalanche paths and landslides.

During the course of our research and database development, several inaccuracies have been detected in published map depictions of lake and riparian habitats. Several such inaccuracies have been revealed through field-work (Butler, 1989d; Butler and Schipke, 1992), whereas others have appeared through combined analysis involving satellite image interpretation, fieldwork, and GIS analysis (Walsh *et al.*, 1989; Butler *et al.*, 1991b).

Scale dependencies and spatial analysis

Fractal and semivariogram analyses

One of the central issues related to the success of using the GIS for bio-physical studies involves the scale dependencies of geographical data and the analysis of multi-scale remote sensing information (Davis *et al.*, 1991). Bian and Walsh (1993) evaluated the scale dependencies between vegetation and topography within Glacier National Park through the use of fractal, semivariance, and multiple regression analyses and the GIS database. The objectives of the research were: to define the effective range of spatial scales within which the Landsat TM-based Reflectance/Absorbtance vegetation index of plant biomass (Frank, 1988) and DEM-based measures of elevation, slope angle, and slope aspect were spatially dependent; to define the degree of the spatial dependencies; and to identify the optimum spatial scale for representing vegetation and terrain relationships within the study area.

> Spatial scale is inherently involved in recognizing spatial patterns on the landscape and in estimating the relationships between landscape components and environmental processes and factors affecting those patterns. Specific biotic, environmental, and historical processes function at various ranges of spatial scales. The ranges vary and overlap among processes and factors. Spatial patterns of the landscape may be discernible at certain spatial scales. The landscape may appear homogeneous at some scales but heterogeneous at others (Nellis and Briggs, 1989). Based upon the assumption that the spatial patterns of landscapes are formed by environmental processes and factors whose effects vary with spatial scale, it is reasonable to hypothesize that the relationships between landscape pattern and the operative processes and factors forming that landscape depend on spatial scale (Meentemeyer, 1989; Turner *et al.*, 1989).
>
> (Bian and Walsh, 1993, p. 1)

Bian and Walsh (1993) and Walsh *et al.* (1991) used semivariograms to establish the range of spatial scale in which spatial dependencies occurred,

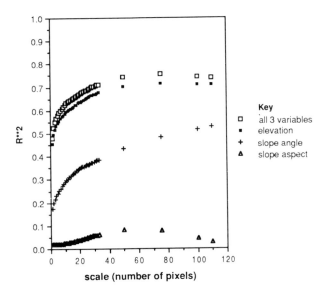

Figure 10.4 Multiple regression values (R²) of the relationship between topography and plant productivity plotted against the spatial scale of data aggregation.

fractal dimensions to identify the strength of the spatial dependencies, and multiple regression analysis to validate the optimum spatial scale, identified by the semivariogram, for explaining vegetation and topographic variation. In general, the research indicated that a characteristic scale existed for the study area in the Park at intervals of approximately 75 pixels (a pixel was 30 m × 30 m). At this scale, variation in vegetation biomass is controlled by the glacial landscape and the systematic arrangement of ridge and valley landforms. Strong spatial dependencies occurred at spatial scales finer than 75 pixel intervals. The semivariogram, which plots semivariance against sampling interval, reached a maximum at the characteristic scale interval and became spatially independent beyond that break in slope. Varying in a similar manner, the relationship between vegetation biomass and topography, represented by R^2 values of multiple regression analysis, were lower at fine scales, reached the maximum at the characteristic scale, and levelled off beyond that scale. Figure 10.4 shows how the R^2 values changed with scale, reaching a maximum value of 0·76 at a scale of 75 pixels. The characteristic scale reveals the most fundamental spatial structure of topography in the study area, and is key to explaining the spatial variation of topography in relation to vegetation biomass and to changes in these relationships as a consequence of terrain (Bian and Walsh, 1993).

The identification of the characteristic scale is useful in defining the optimum spatial scale for data aggregation within the GIS to meet study objectives of variable sensitivity and explanatory power, and for identifying causal processes and factors controlling biophysical relationships within alpine environments. The following spatial techniques are available within the GIS: data sampling; spatial interpolation; spatial aggregation; data simulation; and customized data manipulations programmed in the ERDAS toolkit. They are important to the evaluation of scale dependencies of biophysical variables contained within the GIS.

Spatial autocorrelation and analysis

Brown and Walsh (1993) used Moran's I measure of spatial autocorrelation to identify the spatial scales at which data could be sampled in order to adhere to statistical assumptions of data independence and for the generation of hypotheses regarding biophysical relationships and spatial patterns. Spatial autocorrelation is presented as a bias in the data caused by the spatial ordering of information and relationships between variables as a result of spatial location. The GIS was used to provide iterative sampling of the spatial coverages for sampling lag distances and for measuring the spatial autocorrelation between topographic, biophysical, and satellite-derived land cover types at the alpine treeline. The IDRISI software was used for the calculation of Moran's index.

Analysis of the spatial patterns of the biophysical landscape associated with satellite digital classifications of land cover type and/or GIS spatial coverages of biophysical variables, produced at various spatial scales, is important for characterizing the spatial organization and structure of the alpine environment. Payette *et al.* (1989) showed that the response of treeline to climate change may be affected by the nature of the transition (or spatial structure) along the alpine treeline ecotone. Walsh (1993) assessed the spatial structure of krummholz vegetation (stunted and wind sculptured patches of tree-like vegetation) through the digital classification of Landsat TM and SPOT Multispectral and Panchromatic digital data. Contagion (which measures the adjacency of land cover types and indicates whether or not a clustered landscape is evident), dominance indices (which measure the deviation from the maximum possible landscape diversity), and fractal dimensions (which indicate the complexity of patch shapes on the landscape) were used to assess the representation of krummholz vegetation through each of the reconnaissance platforms and sensor systems (Turner, 1990).

The results indicated the comparative spatial structure of multiple krummholz sites and the relative value of different spatial and spectral resolutions of satellite data in mapping the characteristic krummholz vegetation patches. The land cover classifications contained within the GIS were interfaced to the SPAN spatial analysis software (Turner, 1990) to calculate the three selected measures of landscape structure.

Future research will attempt to characterize the spatial structure of the entire alpine treeline ecotone through numeric indices, such as contagion, dominance, and fractal dimensions, for statistical modelling through sets of biophysical variables contained within the GIS. Measures of the spatial organization of the landscape will serve as dependent variables to be regressed against biophysical descriptors contained within the GIS. Such spatial measures are also being used to characterize the spatial structure of vegetation and terrain within the Park by sorting the alpine environment by elevation, slope angle, and slope aspect zones; basins and sub-basins; and windward and leeward sides of ranges. The basic intent is to obtain measures of spatial structure as an indicator of operative processes shaping the landscape and of variable interrelationships.

Models of alpine treeline

Recent research by several individuals in our research programme has concentrated on biophysical analysis and spatial modelling of environmental conditions at alpine treeline (Butler and Malanson, 1989; Walsh and Kelly, 1990c; Bian, 1991; Kelly, 1991; Brown, 1992; Walsh *et al.*, 1992; Bian and Walsh, 1993; Walsh, 1993). Field-based efforts in support of the GIS analysis and remote-sensing interpretations have concentrated on characterizing the biophysical conditions associated with the current position of alpine treeline, in an attempt to discern if the forest-tundra ecotone will be affected by projected global warming. Results to date have established the erosional importance of animals in this delicate environment (Butler, 1992, 1993). Ongoing analysis of soils, photosynthetically active radiation, and tree rings from krummholz will provide additional data on site conditions associated with treeline in a variety of microsites.

Variation in plant productivity is a key discriminator of the alpine environment. Walsh and Kelly (1990c) measured plant productivity within a portion of the Park through vegetation indices. Analysis of the indices suggested the existence of vegetation conditions that are indicative of variations in alpine treeline form and structure. Measures of vegetation productivity varied as a consequence of the growth forms present in alpine environments and locally and regionally controlled geopedoclimatological variations of the landscape. The typical components of alpine treeline include subalpine forest, krummholz patches of stunted and clumped trees, tundra, and bare surfaces of rock, snow, and ice. The spatial expression and organization of these components are subject to variation in their extent and ordering of occurrence due to environmental constraints. Treeline position, pattern, and topology were regionally assessed using the GIS.

The pattern of vegetation at treeline is complex and variable. In Glacier National Park, a continuous gradient along an elevational transect from tall trees, through shorter trees, to krummholz, and then to tundra is rare. Instead, abrupt changes and reversal of trends are apparent. Local site factors (especially those related to the biological productivity and disturbance at a site) that could result in such patterns are probably controlled in some degree by their topology. That is, the position of the site in the landscape will affect the conditions for photosynthesis (especially available solar radiation, temperature, water, and nutrients) and events such as avalanches and fires.

Payette *et al.* (1989) have recently shown that the response of treeline to climate change may be drastically affected by the nature of the transition from trees through krummholz to tundra, and that the spatial extent and character of the treeline transition relates to its sensitivity for vegetation establishment and growth. Because of the sensitivity of this zone to topoclimatic influences, reconstructed vegetation patterns at treeline have been used to make inferences about past climates (Luckman, 1990; Arno and Hammerly, 1984). Similarly, predictions of global-scale warming are being used to make predictions about the future character of the forest-alpine tundra ecotone. The impacts of long-term climatic conditions on the ecotone are determined by the degree to which these ecosystems are affected by, sensitive to, or in a steady state with, topoclimatic conditions.

Elements of climatic stress on alpine vegetation have been examined by a number of researchers (Bamberg and Major, 1968; Wardle, 1968; Habeck, 1969; Daly, 1984; Hansen-Bristow, 1986). In addition, researchers have reported on various important disturbances on the extent and character of alpine vegetation: degree of irradiation and wind exposure (Daly, 1984; Arno and Hammerly, 1984); density and height of ground cover and depth of winter snow cover (Daly, 1984); and elevation and moisture availability (Peet, 1988). Drifting of snow by wind and the persistence of snowpatches into the growing season are also important issues. Disturbances, such as fire; snow avalanches; debris flows; insects; and grazing, can impede the establishment and growth of mature forests and/or destroy an already mature forest. The importance of disturbances depends upon their intensity and frequency. Our research has focused on developing GIS-based models to characterize solar radiation potential, soil moisture potential, snow accumulation potential, and disturbance regimes. Initial research considered snow-avalanche paths, debris flows, historical fires, and snow accumulation and ablation patterns. The following paragraphs describe the integration of GIS and remote sensing for the development of descriptor variables and modelling approaches to determine the position of the alpine treeline ecotone relative to its observed position.

Modelling topoclimatic factors

The topoclimatic influences on the alpine treeline ecotone were examined through the processing of Landsat TM digital data to classify the vegetation throughout the study area and through the manipulation of GIS coverages. DEMs were used in this research to construct topoclimatic factors affecting alpine treeline; integrate these with remotely-sensed data through classifications; and as the graphical base for data integration.

The topographic influence on spectral responses in satellite data has been well documented (Walsh, 1987). The satellite data were initially corrected for atmospheric and geometric distortions of the data through histogram equalization techniques and geo-referencing approaches. In addition, pre-processing of the satellite imagery included a correction for topographic redistribution of solar radiation (Civco, 1989; Yang and Vidal, 1990).

An unsupervised classification was implemented to cluster spectral radiance values within the TM spectral channels. Statistical information from this classification, interpretation of colour infra-red aerial photography, and field data were used to verify and label the spectral classes into vegetation units. The accuracy of the classification was evaluated through use of the confusion matrix and the Kappa statistic approach (Congalton and Mead, 1983). The objective of the classification was to define vegetation communities within the alpine treeline ecotone. The derived vegetation map was transformed from raster to vector format and entered into the GIS as a stand-alone coverage for subsequent integration in the modelling process.

Three maps of geologic features (Ross, 1959; Oelfke, 1984; and Carrara, 1990) were interpreted to aid in the analysis of substrate conditions and surficial deposits at treeline (Brown, 1991, 1992). Mineralogy is primarily related to parent material, mapped by Ross (1959), stability was determined through a combination of surficial materials (Oelfke, 1984; Carrara, 1990), and surface slope derived from the DEM. Carrara's (1990) map shows the broad distribution of movement-prone substrate, and Oelfke's (1984) map provides detail as to the type of movement and allows inference about the rates of movement. These data were combined to indicate regions where the landscape is too unstable to permit soils to form and vegetation to colonize.

The following paragraphs describe the derivation of some of the topoclimatic variables used in conjunction with the land cover information that were calculated through GIS techniques. Derived models were used as descriptor variables for examining the observed pattern and composition of alpine treeline within the Park.

GIS procedures (after Bonan, 1989) were developed to compute the potential direct solar insolation at any grid location in the study area, given elevation, slope angle, and slope aspect information derived from the DEM (Brown, 1991, 1992). As shadowing by adjacent terrain is an important factor influencing potential insolation in the rugged terrain of the Park, an algorithm for evaluating whether a cell is in shadow at any given time was adapted from Dozier *et al.* (1981) and included in the solar estimation algorithm. The resulting GIS layer represented the relative potential for solar radiation at every site throughout the growing season.

Soil moisture potential was estimated using an index of saturation potential developed by Beven and Kirkby (1979). The index was developed in an attempt to model the effects of subsurface throughflow on water accumulation and soil saturation in catchments which do not experience predominant overland flow (Beven, 1978). The programs used to calculate the index

(Wolock, 1988) were adapted to accept raster data represented as DEMs and to output a raster GIS file representing the relative potential for soil saturation (Brown, 1992).

Frank (1988) used a slope aspect index to model the effects of wind on snow redistribution. A similar, but more complex, topographic modelling approach was developed because of the much greater level of terrain complexity in the study area (Brown, 1992). Dikau (1989) and Skidmore (1990) presented sophisticated approaches to modelling slope position using digital elevation data. A measure of topographic position, curvature, was incorporated into an index of topographic position, which has relevance for estimating potential snow accumulation. A program was developed to compute curvature for a location by determining curvature in four cardinal transect directions, and then summing the four curvatures to yield a non-directional curvature value for each location. An index representing the potential for wind-blown snow accumulation was computed as a function of slope aspect (windward/leeward), elevation, and slope curvature.

Bamberg and Major (1968) and Finklin (1986) presented estimates of the lapse rate (rate of change in temperature with elevation) in the Park. Given information about the temperature at lower-elevation climatic stations, it was possible to interpolate temperatures using the DEM data to sites of which no measurements were made. A GIS layer representing expected mean July temperature was constructed through such an interpolation process.

Topoclimatic variables were used to statistically relate ecologically-critical elements of the landscape, represented as separate raster maps within the GIS, to observed vegetation patterns. Four generalized linear models (TEMTREE: Topographical Empirical Model of Treeline) were constructed for the pattern of each treeline component type (closed canopy forest, open canopy forest, meadow/krummholz/tundra, and unvegetated), using the LOGISTIC program in SAS (Brown, 1992). The independent variables included elevation (related to temperature gradients), potential soil moisture, soil radiation potential, snow accumulation potential, and mapped disturbances. In the first model, constructed for comparison purposes, elevation was the only variable. The second model was constructed using the factors which represented topoclimatic surrogates as predictors. The third model included the topoclimatic factors and disturbances; both those related to geomorphic processes and a historical accounting of recent, large fire events. The fourth model included topoclimatic factors, disturbances, and substrate types.

The model results suggested the relative importance of the controls on vegetation patterns. The Kappa statistic of agreement was used to compare the predicted vegetation patterns with the observed patterns (obtained by satellite image classification and field accuracy assessment) in an independent sample. Kappa values were compared using the techniques of Congalton and Mead (1983) and Congalton (1991). These tests indicated that, overall, inclu-

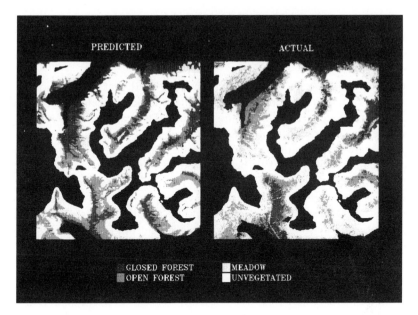

Figure 10.5   Observed pattern of alpine treeline derived from a processed and validated Landsat Thematic Mapper digital classification of vegetation community types, compared with expected pattern of alpine treeline as modelled through LOGISTIC regression of the topoclimatic factors derived through the GIS.

sion of the topoclimatic and disturbance variables in the models resulted in statistically similar patterns between the modelled and observed patterns. Additionally, the models created using elevation as the only variable were significantly less effective than the models which included the other topoclimatic variables and the disturbances. Models which included substrate types were generally less effective.

The responses of vegetation patterns to local topoclimatic processes were modulated significantly by geomorphic and biophysical disturbances. Disturbances from geomorphic processes (e.g. snow avalanches, slope instability, and other talus-producing processes) played a particularly significant role in the treeline vegetation models. The level of explanation of the vegetation patterns was shown to vary geographically, indicating that processes other than those included in the models affected the vegetation patterns.

The best fitting logistic regression equation for each component type, in terms of Kappa values, was used to compute the probability of occurrence of a particular vegetation type at each site. Site probabilities were classified into those expected, and those not expected, to have a particular vegetation type. Expected patterns for each component took the form of new GIS layers and were compared to observed patterns (from the satellite digital classification) through GIS display overlay and summarization operations. Figure 10.5 shows the observed and expected pattern of alpine treeline within an elevation zone from 1600 to 2350 metres. The GIS is well-suited to the spatial and statistical comparison of theoretical and empirical models with observed patterns of phenomena.

The spatial pattern of model errors suggested that, in addition to the local-scale variables studied, drainage basin- to regional-scale factors affected the patterns of vegetation at alpine treeline. The variability in model performance was examined using study area subsets by drainage basin. Subsequent research is examining possible regional and local (basin) controls on alpine treeline which need to be included to improve model specification. Also, spatial autocorrelation in the model residuals, measured using Moran's index, indicated that a spatial process may be operative and critical to understanding the position and character of alpine treeline. Forests, for example, may be more likely to occur at sites which are adjacent to other forested sites (Brown, 1992).

Modelling debris flows

Because local disturbance factors were found to be statistically significant in the TEMTREE models (Brown, 1992), an expanded examination of geomorphic disturbances was begun. Some debris flow sites, once established, experience periodic subsequent disturbances. Some estimate of the frequency and magnitude of these debris flows is important for successful understanding of their effects on treeline. For the debris flows mapped from 1966 aerial photographs, later photographs will be examined to determine if vegetation succession is occurring, indicating a relatively quiescent site without post-1966 debris flows. If additional debris flows are apparent on the more recent photos, this should be apparent by visually assessing the freshness of the deposits in comparison to their 1966 appearance. Freshness is primarily a function of the relative proportions of vegetation and exposed clastic debris.

All young (post-1966) mapped debris flows accessible by trail will be examined in the field. Standard lichenometric techniques will be applied to determine their approximate age of deposition. We anticipate primarily using the lichen *Xanthoria elegans* rather than the more typical *Rhizocarpon geographicum* (Butler, 1982), because of the calcareous nature of the lithology (Osborn and Taylor, 1975; Oelfke and Butler, 1985b). If actual to-the-year dates of tree damage can be obtained from tree-ring analysis, this information can be used to produce a local absolute lichen-growth curve. If such data are not available, the growth curve established by Osborn and Taylor (1975) and successfully applied in the Park by Oelfke and Butler (1985b) will be used with appropriate caution. These data will also allow dating of older, pre-1966 debris flows which have not experienced subsequent mass movement. Spectral signatures from undated debris flows can then be compared to debris flows of known age.

Modelling snowcover

Snow persistence over time and space is a primary control of the pattern of alpine vegetation (Daly, 1984; Peet, 1988). Late-lying snow patches shorten the snow-free period, reducing flowering time, seeding rates, and seedling establishment (Kudo, 1991; Kullman, 1991). Lethal burial of saplings also depresses alpine treeline (Daly, 1984). Multi-temporal Landsat multispectral scanner data throughout a typical snowmelt season within the Park were used to assess the spatial and temporal pattern of snowfield persistence (Allen and Walsh, 1993). Markov chains were used to model changes in the state of snow conditions over time. Seasonal snowmelt can be considered a Markov process because the change in the state of snowcover conditions (full snowcover, partial snow, or no snow) may be influenced by the previous state. The objective of this research was to evaluate the use of Markov transition probabilities and multi-temporal satellite data for estimating the spatial and temporal pattern of spring snowmelt conditions in a rugged and complex mountain environment.

As the alpine ecotone is affected by the rate and pattern of snowmelt, Markov transition matrices can be used to evaluate time and space effects of snow conditions as a biophysical disturbance. Topographically-based sub-matrices, such as for zones of elevation; slope aspect; slope angle; or surface curvature (contained within the GIS), could use a similar approach but with emphasis on environmental controls of vegetation pattern. If Markov models prove as successful as literature and pilot testing have suggested, further analysis could integrate Markov methods with models of solar insolation and soil moisture potential for comparing the interaction of environmental gradients and ecological disturbance. Because of their temporal nature, transition probability matrices can be used to identify areas of relative dynamism whether for snow persistence or for plant succession.

Conclusions

The development of a GIS was an essential element of the spatial and bio-physical research conducted within Glacier National Park. The GIS facilitated the integration of remotely-sensed information with core GIS coverages; GIS coverages and attribute fields derived through manipulation of the core coverages; calculated statistical coverages and spatial measures of feature distance and distribution; and spatially normalized and pattern-defining indices of spatial structure. The spatial operations provided through the GIS permitted not only the verification of remote-sensing inputs to the GIS by comparison to existing GIS coverages and field measurements recorded in the database, but also the validation of GIS coverages by comparison to remote-sensing inputs and to other GIS coverages. The spatial perspective afforded by the GIS also facilitated the generation of new hypotheses and the explanation of established research questions. Spatial operations were essential to the manipulation of the GIS database to support spatial and biophysical queries and for the display of research results as three-dimensional graphics and feature drapes along with statistical findings presented as models, spatial deviations from models, and measures of the spatial pattern, spatial ordering, and spatial structure of diverse phenomena.

Many of the investigations into the Park's biophysical landscape reported in this chapter could not have been conducted without the GIS perspective. The research initiatives presented here represent overarching research objectives that combine to produce an integrated programme on mountain environments that emphasizes spatial and biophysical elements. Each of the projects explores elements applicable to other environments where the analytical procedures can be replicated to examine similar sets of research questions.

The research into spatial dependence and spatial analysis suggests the interrelationship of scale, pattern, and process, as well as the importance of fractal and semivariance analyses to the examination of the characteristic scale that controls much of the primary biophysical variation within the study area. Scale dependencies also have implications for the development of GIS databases because, from both data storage and landscape characterization perspectives, the relatedness of spatial scales in representing biophysical variables is important. Measures of spatial pattern and structure are revealing indicators of biophysical conditions. Such measures indicate the homogeneity, heterogeneity, and dispersion of the landscape, and the resultant perimeter/edge values important in landscape ecology and biogeography. The issue of spatial autocorrelation is important for the development of statistically independent variables suitable for quantitative analyses and for the characterization of biophysical landscapes ordered by spatial position through time.

Our research on alpine treeline provides information on the relationships between multiple and interacting environmental factors and patterns of vegetation. The effects of spatial interactions will be included in subsequent models to test the importance of spatial autocorrelation for describing the pattern of alpine treeline components. Because the spatial pattern of vegetation at an ecotone is an important component of its response to environmental change or disturbance, the determination of the range of patterns and their related processes is important in order to be able to identify which patterns represent potential fronts for the invasion of vegetation as a consequence of global climate change. Models that explore the relationship between the location and character of alpine treeline, together with sets of derived topoclimatic variables that describe the differences between predicted and observed patterns of alpine treeline, were used to develop local disturbance variables to complement the more regional indicators of the treeline ecotone.

Future research into alpine treeline will include the integration of local and regional models and scale-dependent terms. Local process models and regional empirical models will be interfaced and implemented through rule-based models to direct spatial/biophysical queries. Spatial measures of pattern will be explored to characterize ecotone form and interactions within the biophysical landscape. Developments in the GIS database and its structure must be ongoing to support the anticipated research. Issues related to hierarchical data structures for handling scale variation, interfacing models to models and to the GIS, and remote sensing and GIS integration for landscape characterization are among the primary set of considerations related to GIS for mountains.

Acknowledgments

This research was supported through grants from the National Science Foundation (S.J. Walsh and D.R. Butler, PIs), National Aeronautics and Space Administration (S.J. Walsh and D.G. Brown, PIs), and the National Geographic Society (D.R. Butler and G.P. Malanson, PIs). Appreciation is also expressed to the National Park Service, Glacier National Park for providing assistance with a number of research services including housing. Finally, we wish to thank G.P. Malanson; T.R. Allen, Jr.; K.A. Schipke; G. Walsh; W.F. Welsh; H. Hager; and D.M. Cairns for their assistance in this research.

References

Allen, T.R. and Walsh, S.J., 1993, Characterizing multitemporal alpine snowmelt patterns for ecological inferences, *Photogrammetric Engineering and Remote Sensing*, **59**, 10, 1521–29.

Arno, S.F. and Hammerly, R.P., 1984, *Timberline, mountain and Arctic forest frontiers*, Seattle, Washington: The Mountaineers.

Bamberg, S.A. and Major, J., 1968, Ecology of the vegetation and soils associated with calcareous parent materials in three alpine regions of Montana, *Ecological Monographs*, **38**, 2, 127–67.

Beven, K.J., 1978, The hydrological response of headwater and sideslope areas, *Hydrological Sciences Bulletin*, **23**, 4, 419–37.

Beven, K.J. and Kirkby, M.J., 1979, A physically-based variable contributing area model of basin hydrology, *Hydrological Sciences Bulletin*, **24**, 1, 43–69.

Bian, L., 1991, Effects of spatial scale on estimating the relationship between vegetation and topography in a mountainous environment, PhD thesis, University of North Carolina, USA.

Bian, L. and Walsh, S.J., 1993, Scale dependencies of vegetation and topography in a mountainous environment of Montana, *The Professional Geographer*, **45**, 1, 11–1.

Bonan, G.B., 1989, A computer model of the solar radiation, soil moisture, and soil thermal regimes in boreal forests, *Ecological Modelling*, **45**, 275–306.

Brown, D.G., 1991, Topoclimatic models of an alpine environment using digital elevation models within a GIS, *Proceedings, GIS/LIS Conference*, Atlanta, Georgia, USA, 835–44.

Brown, D.G., 1992, Topographical and biophysical modelling of vegetation patterns at alpine treeline, PhD thesis, University of North Carolina, USA.

Brown, D.G. and Walsh, S.J., 1991, Compatibility of non-synchronous in-situ water quality data and remotely-sensed spectral information for assessing lake turbidity levels in complex and inaccessible terrain, *Geocarto International*, **4**, 2, 5–11.

Brown, D.G. and Walsh, S.J., 1992, Relationships between the morphometry of alpine and subalpine basins and remotely-sensed estimates of lake turbidity, Glacier National Park, Montana, *Physical Geography*, **13**, 3, 250–72.

Brown, D.G. and Walsh, S.J., 1993, Spatial autocorrelation in remotely-sensed and GIS data, *Proceedings, American Congress on Surveying and Mapping and American Society for Photogrammetry and Remote Sensing Annual Convention*, New Orleans, Louisiana, USA.

Butler, D.R., 1979, Snow avalanche path terrain and vegetation, Glacier National Park, Montana, *Arctic and Alpine Research*, **11**, 1, 17–31.

Butler, D.R., 1980, Terminal elevations of snow avalanche paths, Glacier National Park, Montana, *Northwest Geology*, **9**, 59–64.

Butler, D.R., 1982, Lichenometric dating in the Mountain Boy cirque, Lemhi Mountains, Idaho, *Journal of the Idaho Academy of Science*, **18**, 1, 15–18.

Butler, D.R., 1983, Observations of historic high magnitude mass movements, Glacier National Park, Montana, *The Mountain Geologist*, **20**, 2, 59–62.

Butler, D.R., 1985, Vegetational and geomorphic change on snow avalanche paths, Glacier National Park, Montana, *Great Basin Naturalist*, **45**, 2, 313–17.

Butler, D.R., 1986a, Snow-avalanche hazards in Glacier National Park, Montana: meteorologic and climatologic aspects, *Physical Geography*, **7**, 1, 72–87.

Butler, D.R., 1986b, Spatial and temporal aspects of the snow avalanche hazard, Glacier National Park, Montana, USA, *Proceedings of the International Snow Science Workshop*, Lake Tahoe, CA, October, 1986, 223–30.

Butler, D.R., 1987, Snow-avalanche hazards, southern Glacier National Park, Montana: the nature of local knowledge and individual responses, *Disasters*, **11**, 3, 214–20.

Butler, D.R., 1988, Teaching natural hazards: the use of snow avalanches in demonstrating and addressing geographic topics and principles, *Journal of Geography*, **87**, 6, 212–25.

Butler, D.R., 1989a, Glacial hazards in Glacier National Park, Montana, USA, *Physical Geography*, **10**, 1, 53–71.

Butler, D.R., 1989b, Canadian landform examples 11 – subalpine snow avalanche slopes, *The Canadian Geographer*, **33**, 3, 269–73.

Butler, D.R., 1989c, Snow-avalanche dams and resultant hazards in Glacier National Park, Montana, *Northwest Science*, **63**, 3, 109–15.

Butler, D.R., 1989d, Geomorphic change or cartographic inaccuracy? A case study using sequential topographic maps, *Surveying and Mapping*, **49**, 2, 67–71.

Butler, D.R., 1990, The geography of rockfall hazards in Glacier National Park, Montana, *The Geographical Bulletin*, **32**, 2, 81–8.

Butler, D.R., 1991, Beavers as agents of biogeomorphic change: a review and suggestions for teaching exercises, *Journal of Geography*, **90**, 5, 210–17.

Butler, D.R., 1992, The grizzly bear as an erosional agent in mountainous terrain, *Zeitschrift fur Geomorphologie*, **36**, 2, 179–89.

Butler, D.R., 1993, The impact of mountain goat migration on unconsolidated slopes in Glacier National Park, Montana, *The Geographical Bulletin*, **35**, 98–106.

Butler, D.R. and Malanson, G.P., 1985a, A history of high-magnitude snow avalanches, southern Glacier National Park, Montana, USA, *Mountain Research and Development*, **5**, 2, 175–82.

Butler, D.R. and Malanson, G.P., 1985b, A reconstruction of snow-avalanche characteristics in Montana, USA, using vegetative indicators, *Journal of Glaciology*, **31**, 108, 185–7.

Butler, D.R. and Malanson, G.P., 1989, Periglacial patterned ground, Waterton-Glacier International Peace Park, Canada and USA, *Zeitschrift für Geomorphologie*, **33**, 1, 43–57.

Butler, D.R. and Malanson, G.P., 1990, Non-equilibrium geomorphic processes and patterns on avalanche paths in the northern Rocky Mountains, USA, *Zeitschrift für Geomorphologie*, **34**, 3, 257–70.

Butler, D.R. and Malanson, G.P., 1992, Effects of terrain on excessive travel distance by snow avalanches, *Northwest Science*, **66**, 2, 77–85.

Butler, D.R. and Malanson, G.P., 1993a, An unusual early-winter flood and its varying geomorphic impact along a subalpine river in the Rocky Mountains of Montana, USA, *Zeitschrift für Geomorphologie*, **37**, 145–55.

Butler, D.R. and Malanson, G.P., 1993b, Characteristics of two landslide-dammed lakes in a glaciated alpine environment, *Limnology and Oceanography*, **38**, 441–5.

Butler, D.R. and Malanson, G.P., 1994, Canadian landform examples – beaver landforms, *The Canadian Geographer*, **38**, 76–79.

Butler, D.R. and Schipke, K.A., 1992, The strange case of the appearing (and disappearing) lakes: the use of sequential topographic maps of Glacier National Park, Montana, *Surveying and Land Information Systems*, **52**, 3, 150–4.

Butler, D.R. and Walsh, S.J., 1990, Lithologic, structural, and topographic influences on snow-avalanche path location, eastern Glacier National Park, Montana, *Annals of the Association of American Geographers*, **80**, 3, 362–78.

Butler, D.R., Oelfke, J.G. and Oelfke, L.A., 1986, Historic rockfall avalanches, northeastern Glacier National Park, Montana, USA, *Mountain Research and Development*, **6**, 3, 261–71.

Butler, D.R., Malanson, G.P. and Oelfke, J.G., 1987, Tree-ring analysis and natural hazard chronologies: minimum sample sizes and index values, *The Professional Geographer*, **39**, 1, 41–7.

Butler, D.R., Malanson, G.P. and Oelfke, J.G., 1991a, Potential catastrophic flooding from landslide-dammed lakes, Glacier National Park, Montana, USA, *Zeitschrift für Geomorphologie* Supplement, **83**, 195–209.

Butler, D.R., Malanson, G.P. and Walsh, S.J., 1991b, Identification of deltaic wetlands at montane finger lakes, Montana, *The Environmental Professional*, **13**, 352–62.

Butler, D.R., Walsh, S.J. and Brown, D.G., 1991c, Three-dimensional displays for natural hazards analysis, using classified Landsat Thematic Mapper digital data and large-scale digital elevation models, *Geocarto International*, **4**, 65–9.

Butler, D.R., Walsh, S.J. and Malanson, G.P., 1991d, GIS applications to the indirect effects of forest fires in mountainous terrain, in Nodvin, S.C. and Waldrop, T.A. (Eds) *Fire and the Environment: Ecological and Cultural Perspectives* Asheville, North Carolina: Southeast Forest Experiment Station, 202–11.

Butler, D.R., Malanson, G.P. and Walsh, S.J., 1992, Snow-avalanche paths: conduits from the periglacial-alpine to the subalpine-depositional zone, in Dixon, J.C. and Abrahams, A.D. (Eds), *Periglacial Geomorphology*, London: John Wiley & Sons, Ltd., 185–202.

Carrara, P.E., 1987, Holocene and latest Pleistocene glacial chronology, Glacier National Park, Montana, *Canadian Journal of Earth Sciences*, **24**, 387–95.

Carrara, P.E., 1990, Surficial geology: Glacier National Park, Montana, *U.S. Geological Survey, Miscellaneous Investigations Series*, MAP I-1508-D.

Civco, D.L., 1989, Topographic normalization of Landsat Thematic Mapper digital imagery, *Photogrammetric Engineering and Remote Sensing*, **55**, 9, 1303–10.

Congalton, R.G., 1991, A review of assessing the accuracy of classification of remotely sensed data, *Remote Sensing of Environment*, **37**, 35–46.

Congalton, R.G. and Mead, R.A., 1983, A quantitative method to test for consistency and correctness in photointerpretation, *Photogrammetric Engineering and Remote Sensing*, **49**, 69–74.

Daly, C., 1984, Snow distribution patterns in the alpine krummholz zone, *Progress in Physical Geography*, **8**, 157–75.

Davis, F.W., Quattrochi, D.A., Ridd, M.K., Lam, N.S-N., Walsh, S.J., Michaelsen, J.C., Franklin, J., Stow, D.A., Johannse, C.J. and Johnston, C.A., 1991, Environmental analysis using integrated GIS and remotely sensed data: some research needs and priorities, *Photogrammetric Engineering and Remote Sensing*, **57**, 6, 689–97.

Dikau, R., 1989, The application of a digital relief model to landform analysis in geomorphology, in Raper, J. (Ed.), *GIS: Three Dimensional Applications in Geographic Information Systems*, New York: Taylor and Francis.

Dozier, J., Bruno, J. and Downey, P., 1981, A faster solution to the horizon problem, *Computers and Geoscience*, **7**, 145–51.

Elias, S.A., 1988, Climatic significance of late Pleistocene insect fossils from Marias Pass, Montana, *Canadian Journal of Earth Sciences*, **25**, 922–6.

Finklin, A.I., 1986, Climatic handbook for Glacier National Park – with data for Waterton Lakes National Park, U.S. Forest Service, General Technical Report, INT-204, Intermountain Research Station, Ogden, Utah.

Frank, T.D., 1988, Mapping dominant vegetation communities in the Colorado Rocky Mountain Front Range with Landsat Thematic Mapper and Digital Terrain Data, *Photogrammetric Engineering and Remote Sensing*, **54**, 12, 1727–34.

Gao, J.K. and Butler, D.R., 1992, Terrain influences on total length of snow-avalanche paths in southern Glacier National Park, Montana, *The Geographical Bulletin*, **34**, 2, 91–101.

Habeck, J.R., 1969, A gradient analysis of a timberline zone at Logan Pass, Glacier National Park, Montana, *Northwest Science*, **43**, 2, 65–73.

Hansen-Bristow, K.J., 1986, Influence of increasing elevation on growth characteristics at timberline, *Canadian Journal of Botany*, **64**, 2517–23.

Kelly, N.M., 1991, Role of topography in the establishment and maintenance of treeline ecosystem components in Glacier National Park, Montana: an integration of remote sensing methods and digital elevation models, Master's Thesis: University of North Carolina, Department of Geography, Chapel Hill.

Kudo, G., 1991, Effects of snow-free period on the phenology of alpine plants inhabiting snow patches, *Arctic and Alpine Research*, **23**, 4, 436–43.

Kullman, L., 1991, Ground frost restrictions of Subarctic Picea abies forest in Northern Sweden: a dendroecological analysis, *Geografiska Annaler*, **73A**, 3–4, 167–78.

Luckman, B.H., 1990, Mountain areas and global change: a view from the Canadian Rockies, *Mountain Research and Development*, **10**, 2, 183–95.

Malanson, G.P. and Butler, D.R., 1984a, Avalanche paths as fuel breaks: implications for fire management, *Journal of Environmental Management*, **19**, 3, 229–38.

Malanson, G.P. and Butler, D.R., 1984b, Transverse pattern of vegetation on avalanche paths in the northern Rocky Mountains, Montana, *Great Basin Naturalist*, **44**, 3, 453–8.

Malanson, G.P. and Butler, D.R., 1985, Ordinations of species and fuel arrays and their use in fire management, *Forest Ecology and Management*, **12**, 1, 65–71.

Malanson, G.P. and Butler, D.R., 1986, Floristic patterns on avalanche paths in the northern Rocky Mountains, USA, *Physical Geography*, **7**, 3, 231–8.

Malanson, G.P. and Butler, D.R., 1990, Woody debris, sediment and riparian vegetation of a subalpine river, Montana, USA, *Arctic and Alpine Research*, **22**, 2, 183–94.

Malanson, G.P. and Butler, D.R., 1991, Floristic variation among gravel bars in a subalpine river, *Arctic and Alpine Research*, **23**, 3, 273–8.

Meentemeyer, V., 1989, Geographical perspectives of space, time, and scale, *Landscape Ecology*, **3**, 3/4, 163–73.

Nellis, M.D. and Briggs, J.M., 1989, The effects of spatial scale on Konza landscape classification using textural analysis, *Landscape Ecology*, **2**, 2, 93–100.

Oelfke, J.G., 1984, 'The location and analysis of landslides along the Lewis Overthrust Fault, Glacier National Park, Montana,' Master's thesis, Department of Geography, Oklahoma State University, Stillwater.

Oelfke, J.G. and Butler, D.R., 1985a, Landslides along the Lewis Overthrust Fault, Glacier National Park, Montana, *The Geographical Bulletin*, **27**, 7–15.

Oelfke, J.G. and Butler, D.R., 1985b, Lichenometric dating of calcareous landslide deposits, Glacier National Park, Montana, *Northwest Geology*, **14**, 7–10.

Osborn, G.D. and Taylor, J., 1975, Lichenometry on calcareous substrates in the Canadian Rockies, *Quaternary Research*, **5**, 111–20.

Payette, S., Filion, L., Delwaide, A., and Begin, C., 1989, Reconstruction of treeline vegetation response to longterm climatic change, *Nature*, **341**, 429–32.

Peet, R.K., 1988, Forests of the Rocky Mountains, in Barbour, M.G. and Billings, W.D. (Eds), *North American Terrestrial Vegetation*, New York: Cambridge University Press, 64–101.

Ross, C.P., 1959, Geology of Glacier National Park and the Flathead Region, northwestern Montana, U.S. Geological Survey, Professional Paper 296, Washington, D.C.: Government Printing Office.

Skidmore, A.K., 1990, Terrain position as mapped from gridded digital elevation model, *International Journal of Geographical Information Systems*, **4**, 1, 33–49.

Turner, M.G., 1990, Landscape change in nine rural counties in Georgia, *Photogrammetric Engineering and Remote Sensing*, **56**, 9, 379–86.

Turner, M.G., Dale, V.H. and Gardner, R.H., 1989, Predicting across scales: theory development and testing. *Landscape Ecology*, **3**, 3/4, 245–52.

Walsh, S.J., 1987, Variability of Landsat MSS spectral responses of forests in relation to stand and site characteristics, *International Journal of Remote Sensing*, **8**, 9, 1289–99.

Walsh, S.J., 1993, Spatial and biophysical analysis of alpine vegetation through Landsat TM and SPOT MX/PAN data, *Proceedings, American Congress on Surveying and Mapping and American Society for Photogrammetry and Remote Sensing Annual Convention*, New Orleans, Louisiana, USA, 426–37.

Walsh, S.J. and Butler, D.R., 1989, Spatial pattern of snow avalanche path location and morphometry, Glacier National Park, Montana, *GIS/LIS '89 Proceedings*, Vol. 1, pp. 286–94.

Walsh, S.J. and Butler, D.R., 1991, Biophysical impacts on the morphological components of snow-avalanche paths, Glacier National Park, Montana, *Proceedings, GIS/LIS Annual Convention*, Atlanta, Georgia, USA, A133–43.

Walsh, S.J. and Kelly, N.M., 1990c, Treeline migration and terrain variability: integration of remote sensing digital enhancements and digital elevation models, *Proceedings, Applied Geography Conference*, **13**, 293–304.

Walsh, S.J., Bian, L., Brown, D.G., Butler, D.R. and Malanson, G.P., 1989, Image enhancement of Landsat Thematic Mapper digital data for terrain evaluation, Glacier National Park, Montana, USA, *Geocarto International*, **3**, 55–8.

Walsh, S.J., Butler, D.R., Brown, D.G., and Bian, L., 1990a, Cartographic modelling of snow-avalanche path location within Glacier National Park, Montana, *Photogrammetric Engineering and Remote Sensing*, **56**, 5, 615–21.

Walsh, S.J., Cooper, J.W., Von Essen, I.E. and Gallager, K.R., 1990b, Image enhancement of Landsat Thematic Mapper data and GIS data integration for evaluation of resource characteristics, *Photogrammetric Engineering and Remote Sensing*, **56**, 8, 1135–41.

Walsh, S.J., Bian, L. and Brown, D.G., 1991, Issues of spatial dependency for surface representation through remote sensing and GIS, *Proceedings, American Congress on Surveying and Mapping and American Society for Photogrammetry and Remote Sensing Annual Convention*, Special Session on The Integration of Remote Sensing and GIS, Baltimore, Maryland, USA, 175–83.

Walsh, S.J., Malanson, G.P. and Butler, D.R., 1992, Alpine treeline in Glacier National Park, Montana, in Janelle, D.G. (Ed.) *Geographical Snapshots of North America*, New York: The Guilford Press, 167–71.

Wardle, P., 1968, Engelmann Spruce at its upper limits on the Front Range, Colorado, *Ecology*, **49**, 3, 483–95.

Wolock, D.M., 1988, Topographical and soil hydraulic control of flow paths and soil contact time: effects on surface water acidification. Doctoral dissertation: Department of Environmental Sciences, University of Virginia.

Yang, C. and Vidal, A., 1990, Combination of digital elevation models with SPOT-1 HRV multispectral imagery for reflectance factor mapping, *Remote Sensing of Environment*, **32**, 35–45.

11

The application of GIS by the National Parks and Wildlife Service of New South Wales, Australia to conservation in mountain environments

Ian Pulsford and Simon Ferrier

Introduction

In recent years, the National Parks and Wildlife Service (NPWS) of New South Wales (NSW), Australia, has increasingly used geographic information systems (GIS) to assist in fulfilling its responsibilities for the conservation and management of mountain environments. This chapter reviews the innovative organizational arrangements for the development and distribution of GIS products within the NPWS, and summarizes a number of case studies of the use of GIS by the NPWS in the conservation of mountain environments in NSW.

Most land management agencies which implement GIS use highly centralized systems. Typically the GIS is based on expensive hardware (mainframe, minicomputer or UNIX work station) located in a central processing or local area network facility. Specialist staff are employed to manage the systems and provide services to clients or management. Some agencies provide for distributed processing (wide area networks) using dedicated land lines or other high-speed data links to remote terminals. These terminals have varying capabilities for remote processing. Generally, the client or staff must contact the centralized specialist or the specialist at the remote node and specify the required output. Often client requests cannot be met expeditiously due to the other priorities of the servicing staff. Communication breakdowns and frustrating delays can occur resulting in inappropriate or untimely outputs to clients. This can reduce interest by non-specialist staff in the use of GIS as a tool to aid land management decisions.

Whilst a centralized approach to the development and distribution of GIS products suits many agencies and scientific institutions, it has proven too costly for some relatively small land management agencies such as the NPWS. This can mean that such an agency either does not use GIS or is forced to make some agreement with a centralized bureau to provide limited services.

This results is users having even less control of data input, analysis, and output.

Decentralized approach to GIS implementation

The organizational structure of the NPWS consists of a head office, five regions, and 25 districts which report to these regions. The NPWS reviewed its GIS requirements in 1989. The organization needed a GIS to: assess the status of natural and cultural conservation in NSW; investigate land acquisition; carry out planning for fire management; control introduced plants and animals in its reserves; protect flora, fauna and cultural resources; assess the environmental impact of the development of visitor facilities; and interpret natural and cultural heritage. The resource attributes needed to provide information for these activities included topography; geology; soils; climate; land tenure; land systems; reserved estate; administrative and political boundaries; vegetation; fire history; roads; tracks and trails; logging history; watercourses; facilities; and records of plants, animals, introduced species, and cultural heritage.

The organization determined that a GIS must be based on the special needs of this essentially field-based organization which has a small computing budget, a generally low level of computing skills amongst field staff, few computing specialists, and the need for all levels of staff to have access to large volumes of spatial environmental data on the natural and cultural heritage of the state. A GIS for this organization must be low-cost and run on existing standard IBM-compatible micro-computer hardware. The system must be easy to use but capable of carrying out complex operations on environmental resource data involving integration, analysis, extrapolation, and reporting. Further, the system must be capable of handling a wide variety of resource attributes for geographical areas ranging in size from small reserves to the whole state ($803\,000\ km^2$). While a search of current GIS software revealed that no suitable commercial system was available for this combination of requirements, a GIS software package known as the Environmental-Resource Mapping System (E-RMS) (Ferrier, 1992c) was suitable. This software was developed in 1987 and was in heavy use in the northern region during 1989. It was then adopted by the whole NPWS in 1989. It has subsequently been marketed commercially and is now licensed by over 80 organizations.

E-RMS is a grid-cell-based GIS for MS-DOS microcomputers. The user interface is entirely menu-driven, and sophisticated data storage and processing techniques enable efficient processing of very large databases on inexpensive hardware. The system uses EGA or VGA graphics and outputs to both cheap colour printers and more expensive plotters. The package includes five modules, for database specification, data import/export (for data exchange), map digitizing, map display/analysis, and predictive modelling.

The unusual aspect of the implementation of GIS in the NPWS is its organization. GIS development and database construction are carried out mostly in four regional centres and, to a lesser extent, head office. In each region, a regional coordinator supervises various staff, including technical digitizing staff and consultants. Databases are constructed by the coordinators and distributed to head office and each administrative unit or district within their region. Coordination of data capture priorities, standards, and database design involves a series of quarterly workshops and meetings as well as frequent exchanges by fax, phone, modem, and mail. As regional coverages are completed, these are forwarded to head office for inclusion in state-wide databases. Some data acquired under licence or data exchange agreement from other agencies are distributed centrally to regions for inclusion in regional databases.

Field staff, such as rangers in district offices and members of special project teams, prepare data and maps and forward them to regional coordinators for digitizing and rapid return as updated databases. Information about the nature and standards of the environmental data captured is stored in regional text-based database management systems. Training of field staff in the use of GIS is carried out in a series of small informal field-based GIS training sessions. These sessions are often used to further evaluate each district's individual GIS needs and to establish a local contact officer for ongoing local liaison. Accessibility by all field staff and local feedback are the major advantages of the NPWS GIS programme. Such feedback has enhanced the process of setting priorities to meet GIS needs at state, regional, and district levels. GIS was introduced into over 40 district offices, regional offices, and head office over a two-year period.

Major projects, such as the broad-scale mapping of vegetation for the whole eastern division of NSW using LANDSAT Thematic Mapper (TM) images, are carefully coordinated through a series of special state-wide workshops. This system for coordinating GIS development depends greatly on a spirit of co-operation. To date, this approach has been very successful. The NPWS would probably not have met its GIS needs as quickly if it had adopted a conventional centralized approach.

An important component of the NPWS' strategy is the purchase, sharing, and exchange of data with other agencies. This involves a number of different arrangements depending on circumstances. Arrangements range from data purchased under licence agreements to memoranda of understanding where data are exchanged, usually without charge. NPWS policy permits exchange of data only where the Service or another agency is the primary custodian of the data to be exchanged. The E-RMS import/export module permits exchange of data with software packages utilized by other agencies. For example, in several collaborative projects with other agencies, outputs from environmental modelling packages, such as BIOCLIM (Busby, 1991), and ARC/INFO GIS databases have been imported into E-RMS databases for further analysis of the data. For example, Norton and Saxon (1992) imported a range of environmental and bioclimatic predictor variables from BIOCLIM and the MEANGROW model (Nix *et al.*, 1992) into NPWS E-RMS databases for further analysis and presentation of mapped output. Agencies such as the Victorian Department of Conservation and Environment are using E-RMS as an easy-to-use alternative to their mainframe-based ARC/INFO GIS systems. This agency is experimenting with the use of ARC/INFO to build primary environmental variables and their transfer to E-RMS databases for use by field staff (F. King, pers. comm.).

The headwaters of NSW – the Great Dividing Range

Approximately 78 per cent of the landscape of NSW is low, flat, or gently undulating land under 300–400 m in elevation (Figure 11.1). The major headwaters in NSW arise along the Great Dividing Range. This somewhat misnamed range, which extends more or less continuously for about 4000 km from southern Victoria, through NSW to Cape York in far northern Queensland, varies in width and height from a series of low narrow ridges to broad plateau uplands of gently tilted or level blocks. In places these plateaux and ranges are deeply incised by gorges. These uplands can be defined as land above 400–500 m in elevation. They occupy about 176 000 km^2 (22 per cent) of NSW.

The Divide forms the watershed for large rivers which flow westward and inland through gradational eucalypt woodlands and mallee into a semi-arid to arid interior. The lands are used for grazing sheep and cattle and for cereal production. Most of these streams flow eventually into Australia's longest river, the Murray-Darling, which eventually enters the sea in South Australia. The streams which flow east of the Divide traverse a relatively narrow coastal plain which varies from 6 km to over 100 km in width.

This coastal plain is the most densely populated part of Australia and contains small areas of dense closed forest (rainforest), eucalypt-dominated communities such as tall open forests, open forests, and woodlands, as well as grassland and coastal heath and swamp complexes. Large areas of the plain have been cleared for agricultural and urban purposes. The southern end of the Divide in NSW is dominated by the Australian Alps where Australia's highest peak, Mount Kosciusko (2228 m), lies at the southern end of the Snowy Mountains. The Alps contain a small area of alpine vegetation (320 km^2) above a treeline of snow gums (*Eucalyptus pauciflora*) at about 1820 m. The subalpine is dominated by snow gums. At lower elevations, there are dense moist tall open mixed eucalypt forests and drier open forests and woodlands.

In many areas, the escarpment and below-escarpment vegetation has not been cleared. A high proportion of the forests and woodlands on the Divide are either protected in a chain of national parks or are within state forests

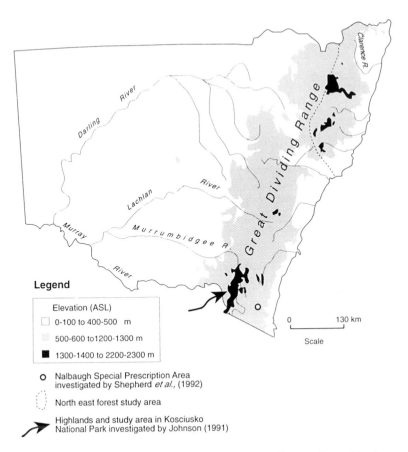

Legend

Elevation (ASL)

☐ 0-100 to 400-500 m

▦ 500-600 to 1200-1300 m

■ 1300-1400 to 2200-2300 m

○ Nalbaugh Special Prescription Area
investigated by Shepherd *et al.*, (1992)

◌ North east forest study area

➤ Highlands and study area in Kosciusko
National Park investigated by Johnson (1991)

0   130 km

Scale

Figure 11.1 Location of studies reviewed in this chapter. They are located in the eastern highlands of NSW along the watershed known as the Great Dividing Range. These highlands are the headwaters of some of Australia's most significant rivers and contain a rich diversity of natural environments and wildlife. (Elevation data courtesy of AUSLIG)

managed for timber production, or in other crown lands. The major threats to the environments in this headwater region vary enormously along its length. In general, these major threats include continuing land clearing, particularly on the western side and on the inland areas of the coastal plain, and logging operations, which are now entering the last available old-growth forests. Other anthropogenic threats include irrigation and electricity generation dams, ski resort development, grazing, mining, mineral exploration, urban development, water pollution, use of recreational four-wheel-drive vehicles, collection of rare plants and animals, invasion by introduced plants, predation of native wildlife by introduced animals, and proposed cloud seeding experiments. Other threats include major natural events, such as bush fires, floods, landslides and, possibly, climate change.

Applications of GIS for conservation in the mountains of NSW

The NPWS has statutory responsibilities for the conservation of flora and fauna throughout the state. One objective of the GIS programme is to provide tools which assist managers to identify and model solutions to problems threatening the conservation of the natural environment both on and off National Parks. The NPWS is therefore developing a GIS for the whole state, as well as for smaller administrative areas.

Examples from daily land management activities

Since GIS is available in all NPWS offices and is easy to use, many staff use it daily. The number of applications of GIS is large and growing as experience increases. They range from simple identification, such as the location and distribution of wildlife; vegetation; geology; terrain; geological sites of significance; electoral and local government boundaries; rare plants; and cultural sites, to assisting in the identification of threats to conservation; the assessment of the conservation potential of natural areas; and more complex environmental modelling procedures. The range of applications of GIS available to NPWS managers is expected to increase substantially as more staff are trained in its use and the number of database coverages increases.

Fire management

GIS is used as an aid during fire suppression operations and as an analytical and presentation tool in fire management planning. It is used to identify and integrate a wide variety of environmental variables, such as vegetation; terrain; slope; and aspect, which have a bearing on fire behaviour, as well as natural and man-made fire barriers such as rivers; cleared land; and fire trails. Fire controllers and strategists can rapidly identify the location of important cultural and natural values which require protection during fire suppression. Fire controllers have access to these GIS data at remote fire control bases, on portable laptop computers and at fire control centres, on desktop personal computers.

GIS is used foremost as a fire recording system. The fire record includes mapping of wild (unplanned) fires and planned fires for fuel hazard reduction, as well as point data indicating the origin and cause of unplanned fires. As the fire record accumulates, GIS will become an increasingly useful tool for the analysis of the history, distribution, and cause of fires. GIS is also used

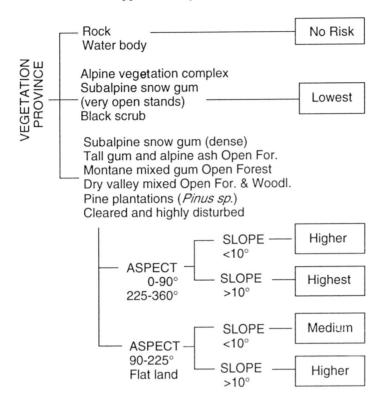

Figure 11.2   Decision-tree model used to derive fire hazard potential map illustrated in Plate 8.

as a fire management planning tool. The E-RMS map display/analysis module permits the rapid combination of variables by boolean overlaying, and the predictive modelling module provides a tool for rapidly building decision-tree models which can then be applied to the database. In one procedure, the fire hazard potential for a reserve is modelled by incorporating expert knowledge with variables such as vegetation type, aspect, and slope to derive a fire hazard potential assessment map (Plate 8 and Figure 11.2). This type of map is used during preparation of fire management plans for reserves.

Environmental assessment

Environmental assessment is carried out on all development proposals within National Parks to fulfil the requirements of the Environmental Planning and Assessment Act, 1979. GIS is used routinely to provide a wide variety of descriptive spatial data, such as the location of rare plants, aboriginal sites, rare faunal habitats, vegetation and other landscape features. For these purposes, GIS is mainly used as a presentation tool.

Rare species

The NSW Wildlife Atlas and the Rare or Threatened Australian Plant (ROTAP) databases are relational text databases which, respectively, store point locality data on fauna and rare and threatened plants. Data from these databases are transferred into E-RMS databases, permitting compilation of inventories and maps of the distribution of common and rare or threatened species for individual parks, districts, or regions. GIS is used routinely to identify locations of rare species during fire suppression operations, environmental assessment, and many other park planning and management activities. The NPWS is experimenting with the fitting of climate surface models, bioclimatic predictors (Nix *et al.*, 1992), to digital terrain models. This will enable the NPWS to produce improved predictive distribution maps for individual species.

Introduced species control

Data from special surveys of introduced species are increasingly stored and analysed using the GIS. For example, a recent study of the population sizes and distribution of broom (*Cytisus scoparius*) in Kosciusko National Park, as part of a wider study in alpine parks, has revealed a greater than 20 per cent reduction of this species as a result of control measures. GIS has helped identify successes in the reduction of this noxious species (Fallavollita and Norris, 1992). Good survey data, combined with the good presentation systems offered by the GIS, have allowed improved strategic planning for ongoing control.

Wilderness assessment

The NPWS is involved in the assessment and management of wilderness under the NSW Wilderness Act, 1987. A number of methodologies have been developed for the assessment of the quality and extent of areas containing wilderness values (Helman *et al.*, 1976; Lesslie *et al.*, 1987). Recently, computer-based techniques have been applied in wilderness assessment to evaluate and map anthropocentric concepts such as naturalness (Lesslie *et al.*, 1987). Whilst these procedures are often qualitative and prone to inconsistency and ambiguity (National Parks and Wildlife Service, 1991), the NPWS has found that GIS techniques are useful as one of several inputs to the subjective assessment of the condition of wilderness, when using the categories of naturalness developed by Laut *et al.* (1977). During these assessments, simple analyses of proximity from disturbance features, such as roads; trails; forest logging; transmission lines; cleared lands; and huts, combined with the perusal of LANDSAT images, give the assessment team an overall impression of the naturalness of the nominated area. These GIS techniques are particularly useful in highlighting those areas which contain the least disturbance.

Research

NPWS personnel conduct research into flora and fauna both independently and in collaboration with external research agencies. GIS is being used in a number of research projects. These include: the design and selection of conservation reserves (Pressey and Nicholls, 1991); predictive mapping and modelling of vegetation (D.A. Keith, pers. comm.); inventory and survey of rare plants and plants which need special attention (Keith and Ashby, 1992); and investigation of the dynamics of inverted treelines and micro-meteorological characteristics of cold air drainage basins or frost hollows in Kosciusko National Park (J. Banks and C. Trevitt, pers. comm.).

Case studies of natural and cultural heritage projects

By legislative mandate, the NPWS is involved in the ongoing issue of wildlife conservation and management not just in its conservation reserves but throughout NSW. Indeed, the most pressing wildlife conservation issues mostly occur outside reserves. A major problem for the NPWS, as with all land management agencies, is the adequacy of environmental data with which to assess the impact of various land management activities or environmental perturbations. The collection and analysis of environmental data, especially biological data, have been notorious for their inadequacies in both the design of field sampling and the analysis of data (Margules and Austin, 1991). Whilst still in its infancy, GIS is being used by the NPWS to revolutionise techniques for the design and analysis of faunal surveys (Ferrier and Smith, 1990; Ferrier, 1992a). GIS has also been used to develop new approaches to predicting wildlife habitat when inadequate survey data on wildlife distribution are unavailable (Shepherd *et al.*, 1992). Examples of the application of GIS to these conservation issues are presented below.

Northeast forests fauna and flora surveys

GIS is playing an integral role in the design and analysis of fauna and flora surveys in the northeast forests of NSW (Ferrier and Smith, 1990; Ferrier, 1991; Ferrier, 1992a). The surveys are providing data for an assessment of the adequacy of regional conservation and future conservation options. This assessment is attempting to answer questions such as how adequate is the existing conservation reserve network? and where should new conservation reserves be located? The project involves both regional and head office staff, and employs sophisticated computer-based techniques developed by the NPWS and the CSIRO (Pressey and Nicholls, 1991, Bedward *et al.*, 1992).

The region being assessed covers approximately 80 000 km². As ground surveys of fauna and flora are expensive and time-consuming, the NPWS

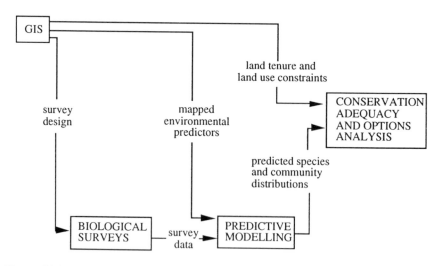

Figure 11.3   Application of GIS to a regional conservation analysis in northeastern NSW.

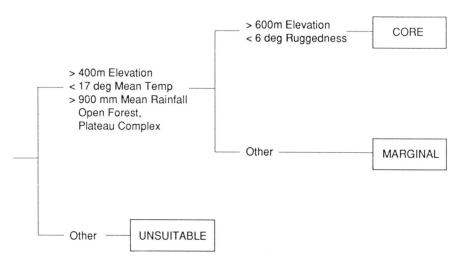

Figure 11.4   Decision-tree model used to derive the predicted habitat for the Hastings River Mouse illustrated in Plate 10.

cannot afford to directly survey every hectare of forest in this region. To overcome this problem, surveys are being conducted at a representative sample of sites scattered throughout the region. Results of these surveys are then interpolated or extrapolated across unsurveyed areas of forest.

The role of GIS in the northeast forests surveys is depicted in Figures 11.3–4 and Plates 9–10. GIS is being used in three ways:

1. to optimize the efficiency of the ground survey design, using the GIS to stratify the location of survey sites in relation to mapped environmental variables (Plate 9);
2. to minimize travel time by locating sites close to access routes; and
3. to analyse correlations between survey results and mapped environmental variables. Predictive models derived from this analysis are then used to interpolate species and community distributions throughout the entire region.

For example, Plate 10 and the decision-tree model in Figure 11.4 illustrate interpolation of the potential distribution of the rare Hastings River Mouse (*Pseudomys oralis*) from environmental data from field survey in the northeast forests study area. In order to assess existing conservation adequacy and future conservation options and priorities, interpolated species and community distributions are analysed in relation to the distribution of conservation reserves and other land tenures.

The GIS database for the northeast forests has a grid resolution of 4 ha, and contains information on terrain, climate, geology, vegetation, disturbance history, land tenure, and access. A digital elevation model (DEM) was fitted to digitized contours and spot heights using software developed by Hutchinson (1989). The DEM was used to derive a number of secondary terrain attributes such as slope, aspect, topographic position, ruggedness, and exposure to the northern sky. The DEM was also used to generate a series of climatic surfaces from models developed at the Centre for Resource and Environmental Studies (Hutchinson, 1987). General vegetation systems were mapped for the entire region from LANDSAT TM imagery (Roberts, 1992). Detailed vegetation association mapping has also been incorporated into the database where available.

For designing field surveys, the GIS was used to generate maps depicting different combinations of environmental variables. For example, the fauna survey sites were stratified according to generated combinations of vegetation (five classes), elevation (three classes), geology (seven classes) and topographic position (three classes). Of the 315 possible combinations, 201 occur in the region. In the 1991/92 financial year, fauna surveys were conducted at 380 sites scattered throughout these survey strata. A further 270 sites will be surveyed in 1992/93.

A range of techniques is being used to analyse and model relationships between the distributional data for each species surveyed and the mapped environmental variables stored in the GIS. These techniques include simple profile or homoclime matching (Busby, 1991), generalized linear modelling (Austin *et al.*, 1984), and decision-tree induction (Stockwell *et al.*, 1990). Decision-tree induction tools have been recently incorporated into the E-RMS package in the form of a Predictive Modelling Module (Ferrier, 1992b). This module facilitates efficient production of decision-tree models and interpolation of these models across any geographical area of interest.

*Guidelines for logging old-growth eucalypt forest to
maximize conservation of wildlife*

The NSW Forestry Commission's continued logging of old-growth and, to a
lesser extent, secondary forests is highly controversial, as it threatens the
habitats of rare and endangered wildlife (Shepherd *et al.*, 1992). Old-growth
forests are the last remaining unlogged sclerophyll and rainforest areas in the
State. These forests occur in headwaters high on the eastern side of the Great
Dividing Range. The sclerophyll forests which provide the most extensive
habitat for wildlife are dominated by eucalypts, with varying mesic to dry
understorys of small trees, shrubs and forbs. In some places, rainforest occurs
along creek lines in eucalypt forest. Wildlife in these forests which is most
threatened by logging operations includes birds, mammals nesting in arboreal
hollows (including gliding possums), some terrestrial mammals, owls, reptiles,
and amphibians.

In recent years, the development of a land use conflict in southeastern
NSW has caused the NSW Government to establish two Special Prescription
Areas (SPAs). These were to be used for timber extraction, but with modification of harvesting practices to minimize disruption of wildlife corridors
which link three proposed national parks (Shepherd *et al.*, 1992) (Figure 11.1).
To address the NPWS' concerns over the effect of logging on the conservation
of wildlife in these areas, a special team of NPWS staff and independent
wildlife ecologists was formed. While there was limited information on the
distribution of flora and fauna, some data, such as forest type and topography, were available. Due to the extremely short time-frame for a response to
government, a novel approach to modelling habitat areas for wildlife using
expert knowledge and available environmental data was needed. GIS was
chosen as the best tool to provide rapid analysis of available spatial environmental data. It also provided a valuable means to graphically illustrate
how proposed options for a retained habitat network would affect existing
logging schedules.

To assist decision-making, eight guiding ecological principles were
established prior to commencing analysis of the environmental data. These
guidelines included concepts such as the necessity to place the requirements
of wildlife in a regional context, the need to minimize habitat fragmentation
and ensure linkages between important forest habitats, the need for proper
planning before logging or road construction, recognition of the conservation
value of old-growth forest, the need to design viable conservation reserves,
and the principle that a conservative approach to forest management should
be adopted until more data on wildlife are available.

To meet the needs of the guiding ecological principles, the establishment
of Population Assisting Links (PALs) within the 1270 ha Nalbaugh SPA was
proposed. PALs were defined as relatively linear areas of retained forest of
high habitat quality which are large enough to allow for the dispersal of
species between fragments, to support resident populations of key fauna over

the long term, and adequately buffered against edge effects. The minimum suggested width of a PAL was one km, but should never be less than 700 m. PALs connected the proposed new National Parks, as well as areas reserved from logging within the Nalbaugh SPA.

In order to derive PAL options, a special purpose database with 50 m × 50 m grid cells was assembled and analysed using the E-RMS package. The data used in this analysis included elevation; slope; aspect; streams; roads; trails; fire history; vegetation types; and fauna records, as well as the boundaries of reserves, proposed National Parks, logging coupes, and the Nalbaugh SPA. The E-RMS map display/analysis module was used to derive relative habitat value maps of high spatial resolution for a range of vertebrate fauna in the Nalbaugh SPA. The stages in this process were as follows:

1. the dominant forest type and topographic domain of each grid cell were identified;
2. major divisions of topographic positions (flat and lower slopes, midslopes, or ridge) of slopes were drawn by eye on a topographic base and digitized (computed approaches to this variable did not produce a better result);
3. the two ridge top classes were merged as it was considered that aspect was not relevant on the ridge tops; and
4. by overlaying the topographic position and a simplified aspect variable (two aspects) five topographic domains were defined.

Finally, the 16 forest types were given a score, based on the estimated value of each forest type as habitat for arboreal marsupials, owls, terrestrial mammals, and diurnal birds. These scores were adjusted according to topographic domain category, to account for variations in habitat quality due to topography. This approach generated a matrix of 80 relative habitat quality scores (16 vegetation types by five topographic domain categories). Any cells within the matrix with the same score were merged within E-RMS to produce a map of relative habitat quality.

The resulting habitat map was used to design a PAL between the proposed reserves (Plate 11). Five options for configurations of the PAL were examined in terms of their costs and benefits for wildlife and wood production. Each option was assessed in terms of its considered impact on existing logging schedules and whether it met the guidelines for establishing PALs to connect areas of highest-quality wildlife habitat. All options centred on the main watercourse and encompassed the high-quality moister forest types. The GIS was able to provide detailed breakdowns of the areal extent of various forest planning units, such as production forest and reserved areas, of each PAL design. Only three of the five PALs presented were considered to meet the acceptable criteria established by the team. There was no design which would not affect proposed logging schedules in some way.

This case study illustrates that GIS, together with the input of expert knowledge from a team of wildlife ecologists and park planners, can be used to produce a highly detailed spatial analysis of habitat values for wildlife. Existing data were assembled in the database in one week. In one month, a powerful case was developed for the design of wildlife corridor options to link a system of conservation reserves and yet still provide for the political imperative for some logging to occur. The E-RMS also provided an effective tool for the presentation of the outcomes of the analysis to the decision-makers. It is clear that such approaches, however sophisticated the procedure and technology, will never replace the collection of adequate biological data through extensive wildlife survey and research. Such modelling exercises need to be thoroughly validated by detailed wildlife surveys during forest harvesting.

Cultural heritage conservation

The NPWS is the statutory body responsible for the protection and conservation of Aboriginal archaeological, mythological, and sacred sites throughout NSW, and maintains extensive computer- and map-based registers which record the location and characteristics of these sites. These data are now being incorporated into GIS databases to assist archaeologists to relocate, conserve, interpret, and investigate the nature and distribution of sites. Johnson (1991) conducted a major survey of sites in Kosciusko National Park in the Great Dividing Range, one of the state's largest parks (6600 km²). GIS was used extensively in this study to analyse and present data on the occurrence, distribution, and type of sites recorded before and during the study. The concept of archaeographic regions was first proposed by Turner (1991) and was used by Johnson (1991) to broadly stratify locations for archaeological field surveys. A slope map from a database with 200 m × 200 m grid cells was used to assist a panel of archaeologists and staff with expert local knowledge in defining archaeographic regions. Following the archaeological field surveys, all existing and new data were combined and the GIS was used to calculate the frequency of sites in each elevation class. This revealed distributions of sites relating to several distinct environmental zones. Other environmental variables such as aspect and slope were analysed in a similar way.

More detailed analysis of terrain data was not possible due to the relative coarseness of the terrain models and the lack of adequate stream data at the smallest available map scale (Johnson, 1991). More sensitive analysis of the environmental parameters associated with the pattern of past human use of this landscape would require a more refined digital terrain model, a landscape unit map, and environmental models describing the predicted distribution of vegetation during past climates (M. Sullivan, pers. comm.).

Conclusions

The use of GIS by the NPWS for the conservation of mountain environments is still relatively new. Database construction and the production of the

E-RMS package have only been in progress since 1986. The ease of use, speed, and availability of the package in all NPWS offices has resulted in rapid acceptance by many field staff. The full potential of the system is yet to be explored; this will occur more rapidly with the recent release of the new modelling module for the E-RMS package. This wildlife conservation and land management agency is using GIS in an ever-increasing number of ways which permit it to efficiently provide information and analyses of large volumes of environmental data to the government. These data can be assembled in relatively short periods and provide a vital tool on which to base decisions. GIS technology greatly improves the Service's ability to carry out its responsibilities for conservation. While the E-RMS package may eventually be replaced by other software products, GIS is likely to remain indispensable to the NPWS in its task of conserving the flora and fauna of the state.

Acknowledgments

The assistance and support of Regional Manager Graeme Worboys and Deputy Director (Field Management) Alistair Howard for this work is gratefully acknowledged. Tim Shepherd and Andy Spate kindly reviewed drafts and provided helpful comments. Tim Shepherd kindly helped prepare a figure. The considerable assistance of Eda Addicott, Malcolm Stephens, Murray Ellis, Anne Blaxland, Mike Saxon, Michael Bedward, David Keith, Nick Gellie, Steve Naven, Genevieve Wright, Elissa Fallavollita, Mike Young, departed digitizing staff and software programmers throughout the GIS development programme is gratefully acknowledged. The Australian National Parks and Wildlife Service provided considerable funding for software development. We thank the Drs Tony Norton, Hugh Possingham and David Lindenmeyer for generously allowing us to cite unpublished work on guiding ecological principles.

References

Austin, M.P., Cunningham, R.B. and Fleming, P. M., 1984, New approaches to direct gradient analysis using environmental scalars and statistical curve-fitting processes, *Vegetatio*, **55**, 11–27.

Bedward, M., Pressey, R.L. and Keith, D.A., 1992, A new approach for selecting fully representative reserve networks: addressing efficiency, reserve design and land suitability with an interactive analysis, *Biological Conservation*, **62**, 115–25.

Busby, J.R., 1991, BIOCLIM – a bioclimate analysis and prediction system, in Margules C.R. and Austin, M.P. (Eds) *Nature Conservation: Cost Effective Biological Surveys and Data Analysis*, Report 64–7, Canberra: CSIRO.

Fallavollita, E. and Norris, K., 1992, 'The occurrence of broom *Cytisus scoparius* in the Australian Alps National Parks,' unpublished report to the Australian Alps National Parks Liaison Committee, Canberra: Australian National Parks and Wildlife Service.

Ferrier, S., 1991, A GIS database for biodiversity conservation evaluation and planning on the NSW North Coast, in *Proceedings of the Conference on Remote Sensing and GIS for Coastal Catchment Management*, Lismore: Centre for Coastal Management, University of New England, 81–8.

Ferrier, S., 1992a, Computer-based spatial extension of forest fauna survey data: current issues, problems and directions, in Lunney, D. (Ed.) *Conservation of Australia's Forest Fauna*, Sydney: Royal Zoological Society of New South Wales, 221–7.

Ferrier, S., 1992b, Development of a Predictive Modelling Module for E-RMS, unpublished report to the Australian National Parks and Wildlife Service, Sydney.

Ferrier, S., 1992c, *Environmental-Resource Mapping System software package*, Sydney/Canberra: New South Wales National Parks and Wildlife Service and Australian National Parks and Wildlife Service.

Ferrier, S. and Smith, A.P., 1990, Using geographical information systems for biological survey design, analysis and extrapolation, *Australian Biologist*, **3**, 105–16.

Helman, P.M., Jones, A.D., Pigram, J.J.J. and Smith, J.M.B., 1976, *Wilderness in Australia. Eastern New South Wales and South Western Queensland*, Armidale: Department of Geography, University of New England.

Hutchinson, M.F., 1987, Methods of generating weather sequences, in Bunting, A.H. (Ed.) *Agricultural Environments: Characterisation, Classification and Mapping*, Wallingford: CAB International, 149–57.

Hutchinson, M.F., 1989, A new method for gridding elevation and stream line data with automatic removal of spurious pits, *Journal of Hydrology*, **106**, 211–32.

Johnson, I., 1991, Kosciusko National Park baseline heritage study (Aboriginal sites), unpublished report to the New South Wales National Parks and Wildlife Service, Sydney.

Keith, D.A. and Ashby, E., 1992, *Vascular plants of conservation significance in the south east forests of New South Wales*, Occasional Paper No. 11, Sydney: New South Wales National Parks and Wildlife Service.

Laut, P., Heyligers, P.C., Keig, G., Laffler, E., Margules, C., Scott, R.M. and Sullivan, M.E., 1977, *Environments of South Australia*, Canberra: CSIRO Division of Land Use Research.

Lesslie, R.G. and Taylor, S.G., 1985, The wilderness concept and its implications for Australian wilderness preservation policy, *Biological Conservation*, **32**, 309–33.

Lesslie, R.G., Mackey, B.G. and Preece, K.M., 1987, A computer based methodology for the survey of wilderness in Australia, consultant's report to the Australian Heritage Commission, Canberra.

Margules, C.R. and Austin, M.P., 1991 (Eds) *Nature Conservation: cost effective biological surveys and data analysis*, Canberra: CSIRO.

National Parks and Wildlife Service, 1991, *Goodradigbee wilderness assessment report*, Sydney: New South Wales National Parks and Wildlife Service.

Nix, H.A., Stein, J.A. and Stein, J.L., 1992, Developing an environmental geographic information system for Tasmania. An application for assessing the potential for hardwood plantation forestry, consultancy report to the Land Resources Division and Bureau of Rural Resources, Canberra: Federal Department of Primary Industries and Energy.

Norton, T.W. and Saxon, M.J., 1992, The potential distribution of the koala in southeast New South Wales and recommendations for its survey and management, consultancy report, Centre for Resource and Environmental Studies, Australian National University, Sydney: New South Wales National Parks and Wildlife Service.

Pressey, R.L. and Nicholls, A.O., 1991, Reserve selection in the Western Division of New South Wales: development of new procedures based on land system mapping, in Margules, C.R. and Austin, M.P. (Eds) *Cost Effective Biological Surveys and Data Analysis*, Canberra: Nature Conservation, CSIRO.

Roberts, G.W., 1992, Vegetation systems of Northeastern New South Wales Mapped from LANDSAT TM Imagery, unpublished report, Sydney: NSW National Parks and Wildlife Service.

Shepherd, T.G., Saxon, M.J., Lindenmayer, D.B., Norton, T.W. and Possingham, H.P., 1992, *A proposed management strategy for the Nalbaugh Special Prescription Area based on guiding ecological principles*, South East Forest Series No. 2., Sydney: NSW National Parks and Wildlife Service.

Stockwell, D.R.B., Davey, S.M., Davis, J.R. and Noble, I.R., 1990, Using induction of decision trees to predict Greater Glider density, *AI Applications*, **4**, 33–43.

Turner, I., 1991, Cultural resources database for the Murray, interim report to the Murray-Darling Basin Commission from the NSW National Parks and Wildlife Service, Sydney.

12

A GIS analysis of forestry/caribou conflicts in the transboundary region of Mount Revelstoke and Glacier National Parks, Canada

Sandra J. Brown, Hans E. Schreier, Guy Woods and Susan Hall

Introduction

The Canadian National Parks Service has identified transboundary land use problems as having a serious impact on park resource management. To evaluate these external influences, comprehensive knowledge is required on the resource values of adjacent land, their importance and interdependence with resources in the national parks, the environmental impact of various developments, and the effect of recreational activities on the regional ecology. The mandates and priorities of agencies sharing responsibility for land management in park boundary areas also need to be identified. Integrated resource management goals can then be developed for larger management units, including both park and adjacent land, based on ecological criteria.

Mount Revelstoke and Glacier National Parks in British Columbia (BC) incorporate approximately half the range required by a minimum viable herd of woodland caribou (Simpson and McLellan, 1990). Habitat quality within the parks is generally poorer than in adjacent drainage areas, due to the high average elevation of the parks and the predominance of semi-mature forest inside them (Simpson and McLellan, 1990). Problems of habitat loss in the areas surrounding the parks have been aggravated by a major transportation corridor and the flooding of key low-elevation habitats as a result of the development of the Mica and Revelstoke hydroelectricity dams. Habitat fragmentation through logging, isolating the parks from seasonal habitats outside their boundaries, would jeopardize the survival of a viable caribou population. Movement corridors (linkages) and the protection of key habitats outside the parks are critical to their functioning as an effective ecological reserve.

This study is in response to concerns about the loss of caribou habitat quality due to intensive forest management in the areas bordering Mount

Revelstoke and Glacier National Parks. The long-term consequences of such habitat loss adjacent to the parks are evaluated through:

1. compilation of a geo-referenced digital database of basic resource information;
2. identification and modelling of critical caribou habitat using a geographic information system (GIS);
3. evaluation of existing land use and long-term harvesting plans; and
4. development of a GIS-based methodology combining the two evaluations, to assist conflict resolution between forestry and wildlife.

The spatial arrangement and seasonal distribution of food and cover play a key role in arriving at the optimum habitat needed to maintain a viable herd of caribou. Habitat requirements for caribou are restrictive due to their dependence on old-growth forest. Low-elevation old growth is required for snow interception, ground forage, lichen litter, and predator avoidance habitat. Lichens in high elevation forests are a primary food source during the summer and the latter part of winter (Stevenson and Hatler, 1985; Simpson and Woods, 1987; Simpson and McLellan, 1990). Contiguous corridors of mature or semi-mature timber are required by caribou as travel routes between seasonal habitats, most importantly from ridge tops to valley floors. Caribou may be unwilling to cross large areas of young seral vegetation, due to high predation risk; isolation of otherwise suitable habitat can reduce the potential of an area to support caribou (Antifeau, 1987; Simpson and McLellan, 1990). The habitat of woodland caribou varies seasonally with forage availability, snow accumulation, snowpack solidification, snowmelt, green-up, and calving. Caribou habitat may be divided seasonally into early winter, late winter, spring, calving and summer periods (Simpson and McLellan, 1990).

The 42 994 ha Gold/Bachelor watershed located north of Glacier National Park was chosen for this study because the drainage basin provides a site where:

1. future development conflicts can be modelled using current forest management scenarios;
2. management modifications can be made before the area is fully developed or utilized;
3. there is sufficient variation in available forest ecology to develop a case study showing how GIS-based models can assist in conflict resolution between park and forestry use.

Methods

Digital database compilation

A digital database for the Gold/Bachelor watershed was compiled from a number of data sources. British Columbia's Terrain Resource Information

Management (TRIM) digital data sets at 1:20 000 scale provide the basemap for the Gold/Bachelor drainage system. The TRIM digital basemap provides information on planimetric position and topography, including: contours; rivers; lakes; marshes; swamps; glaciers; icefields; eskers; moraines; and slides (BC Ministry of Environment and Parks, 1988).

The forest resource inventory data was obtained in digital form from the Inventory Branch, BC Ministry of Forests. The data include the digital forest cover positional (SEG) files at 1:20 000 scale and associated forest inventory attribute (FIP) files. The forest cover positional files divide the land into homogeneous units with similar forestry characteristics. The forest cover attribute files contain information on species, species characteristics (such as height, age, crown closure), site characteristics, ownership, and management activities (Inventory Branch, 1990). The forests that can potentially be used for logging were quantified in two ways. First, an operable forest map was produced which includes all forests that are physically accessible and environmentally sound for harvesting, using currently available technology. Operable forest must also contain economically viable timber, based on the previous business cycle which covers activities from the time of the last highest to lowest profit period. Secondly, a 40-year forest harvest plan was obtained in digital format at 1:20 000 scale from the Golden Forest District, British Columbia Ministry of Forests. The areas to be logged in this plan are within the operable forest, and the forest stands to be harvested were selected in terms of sufficient volume in readily accessible areas that have the least terrain stability problems. Both data sources were incorporated into the GIS database.

Seasonal caribou habitat modelling

A caribou habitat model was developed in co-operation with K. Simpson, Keystone Bio-Research, and B. McLellan. The model was patterned after the Woodland Caribou Cumulative Effects Analysis Model developed by the US Department of Agriculture (Summerfield *et al.*, 1985). The model is based on species-habitat relationships observed by local wildlife experts over 10 years (Simpson and Woods, 1987), and will be refined as new information from radio-transmittance studies are gathered. The model was developed for use within the Revelstoke and Golden Forest Districts, specifically the Gold/Bachelor watershed (Figure 12.1).

Caribou habitat quality and availability within the Mount Revelstoke and Glacier National Parks region are limited during early winter, late winter and spring (Simpson and McLellan, 1990). These three critical seasonal habitats are used in the habitat selection model for the trans-boundary area. Summer and calving habitats were not included in the assessment since there appears to be a sufficient amount of these habitats within the park boundaries.

Topographical and vegetation mapping criteria for caribou are based

NORTH COLUMBIA MOUNTAINS TEST AREA

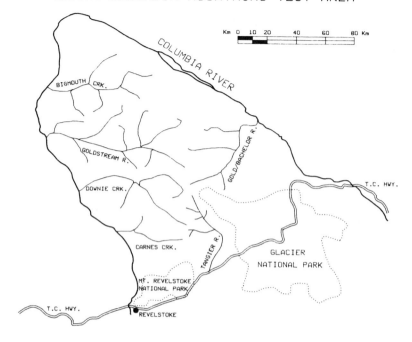

Figure 12.1   Gold/Bachelor watershed.

Table 12.1   Overall habitat suitability ranking

Suitability Class	Suitability (%)	Rank
High	80–100	1·0
Medium	50–80	0·7
Low	15–50	0·3
Nil	0–15	0·0

primarily on Revelstoke radio tracking data (Simpson and McLellan, 1990). Habitat use information was synthesized to rate habitat suitability and scaled to produce an index value between 0 (unsuitable habitat) and 1 (optimal habitat) (Table 12.1). The habitat suitability is based on physiographic and vegetative parameters which differ by season. Thus it is possible to model the habitat suitability for each season through physiographic variables (elevation, slope and slope position) and four vegetation variables (dominant species, height class, crown closure class and riparian habitat). The physiographic information was provided by the TRIM digital data and the vegetation information was extracted from the BC Ministry of Forest's forest cover positional and attribute files.

Table 12.2   Early winter caribou habitat suitability ranking

| Variable | Suitability Class | | | |
	High	Medium	Low	Nil
Elevation (m)	0–1199	1200–1499	1500–1799	≥1800
Slope (%)	<29	30–39	40–69	≥70
Slope position	Lower	Mid valley	Upper	Crest
Dominant species	Spruce Cedar hemlock True fir	–	Douglas fir Lodge pine White bark	Others
Height class	≥3	–	2	1
Crown closure class	6–10	–	4–5	0–3

Table 12.3   Late winter caribou habitat suitability ranking

| Variable | Suitability Class | | | |
	High	Medium	Low	Nil
Elevation (m)	1400–1799	1800–1999	0–1199	≥2000
Slope (%)	0–19	20–39	40–64	≥65
Slope position	Crest Upper	Mid	Valley Lower	–
Dominant species	Spruce True fir	–	Cedar Hemlock	Others
Height class	≥2	–	1	–
Crown closure class	2–6	1	7–9	0

The variables and class limits used to develop habitat suitability evaluations are displayed seasonally in Tables 12.2–4. The classes shown indicate habitat suitability rankings between 0 (nil suitability) and 1 (high suitability). To determine the seasonal habitat suitability, a generalized theme was created within the GIS system for each variable and its respective class boundaries as given in Tables 12.2–4. Each class was ranked according to its suitability. For a given season, the relevant themes were then overlain. The rankings of the individual variables for the resultant polygons were weighted according to Table 12.5, summed and then divided by the total weight for a given season, producing a seasonal rank ranging from 0 to 1. A very low value for any one variable will produce a low suitability ranking for a given polygon. Land not productive for caribou habitat, such as glaciers, icefields, moraines, screes, and slides was placed into a barren land category.

Critical caribou habitat modelling

The three seasonal caribou habitat submodels were combined to assess the overall caribou habitat suitability. The seasonal submodels were generalized

Table 12.4   Spring caribou habitat suitability ranking

| Variable | Suitability Class | | | |
	High	Medium	Low	Nil
Elevation (m)	0–1199	1200–1499	1500–1799	≥1800
Slope (%)	0–19	20–64	65–79	≥80
Slope position	Valley	Low Mid	–	Crest
Dominant species	Others	–	–	Lodge pine Whitebark
Crown closure class	0–1	2–4	5–9	–
Riparian	Yes	–	–	No

Table 12.5   Weighting factors for seasonal caribou habitat suitability

Variable	Early winter	Late winter	Spring
Elevation	4	4	2
Slope	3	3	1
Slope position	4	4	3
Dominant species	3	3	1
Height class	5	4	–
Crown closure	5	2	3
Riparian	–	–	3
Total	24	20	13

and filtered to remove any polygons less than 1 ha in area, and then overlain and summarized using the classes given in Table 12.6. Class I habitat contains all mapping units with high suitability in either early winter, late winter or spring, separately or in combination. Class II habitat contains all mapping units with medium suitability in early winter, late winter and spring combined. Class III habitat contains all mapping units with medium suitability in two of the three seasons. Class IV habitat contains all mapping units with medium suitability in either early winter, late winter or spring separately but not in combination, and all mapping units with low or nil suitability except barren land. Class V habitat contains all mapping units corresponding to barren land.

Habitat/forestry conflict identification

The combined caribou habitat distribution for the winter to spring period was superimposed first onto the operable forest map and then on the 40-year harvesting plan within the operable forest. This provided a spatial analysis of the location and extent of conflicts between timber harvesting and high quality caribou habitat.

Table 12.6   Habitat suitability classes

Class	Early winter		Late winter		Spring
I	High	and/or	High	and/or	High
II	Medium	and	Medium	and	Medium
III	Medium	and	Medium		
	Medium			and	Medium
			Medium	and	Medium
IV	Medium	or	Medium	or	Medium
V	Barren	and/or	Barren	and/or	Barren

Results

The GIS analysis first enabled us to show the extent of early winter (Figure 12.2), late winter, and spring suitable caribou habitats in the 42 994 ha test watershed (Figure 12.3). The seasonal availability of suitable caribou habitat varies dramatically. Spring caribou habitat is most limiting, with only 13 per cent of the watershed in the high and medium suitability classes and 72 per cent of the watershed in the nil and barren suitability classes. Early winter caribou habitat is slightly more abundant with 16 per cent of the watershed in the high and medium suitability classes and 48 per cent in the nil and barren suitability classes. Late winter caribou habitat is the most abundant, with 27 per cent of the watershed in the high and medium suitability classes and only 17 per cent (largely glaciers and icefields) in the nil and barren suitability classes.

The resultant combined caribou habitat suitability map is shown in Figure 12.4. The high suitability habitat classes from the early winter, late winter, and spring submodels are largely non-coincident: only 631 ha (11 per cent) overlap. Class I habitat covers 13 per cent of the watershed and is spread throughout the major sub-basins. The medium suitability classes from the early winter, late winter, and spring submodels are also largely non-coincident, with 14 787 ha (83 per cent) not overlapping. Class II and III habitats are limited, covering only 6 per cent of the watershed, but provide valuable movement corridors between Class I habitats.

The area of each caribou habitat class (I–V) that overlaps with operable forest is summarized in Figures 12.5 and 6. A total of 46 per cent of Class I habitat and 32 per cent of Class IV habitat are coincident with operable forest. Comparison of the 40-year harvest plan with the caribou suitability map indicated that a total of 980 ha is to be harvested, removing 22 per cent of the operable timber. The overlap between caribou habitat and the 40-year harvest plan indicates that 52 per cent of the forest to be harvested is Class I caribou habitat (Figure 12.6). Only 6 per cent of the harvest area is from Class IV habitat, while 32 per cent of this class is operable.

GOLD / BACHELOR WATERSHED

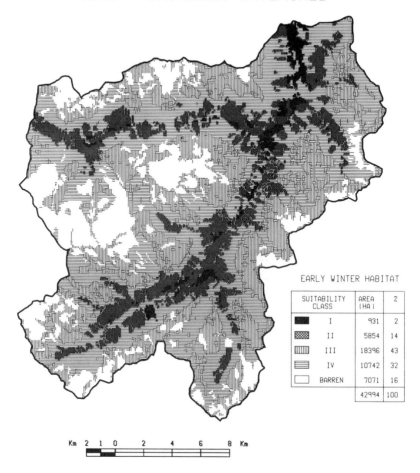

EARLY WINTER HABITAT

SUITABILITY CLASS		AREA (HA)	%
	I	931	2
	II	5854	14
	III	18396	43
	IV	10742	32
	BARREN	7071	16
		42994	100

Km 2 1 0   2   4   6   8 Km

Figure 12.2   Early winter caribou habitat suitability.

Discussion

Woodland caribou populations are declining in British Columbia and are no longer present on many of their former ranges (Bergerud, 1978; Fish and Wildlife Branch, 1979; Simpson and Woods, 1987). Timber harvesting potentially affects caribou in two ways: it alters the structure of the forest, and it creates access to caribou ranges. Because the mountain ecotype caribou is dependent on arboreal (tree-borne) lichens, it is vulnerable when old-growth forests are harvested. Young productive forests are favourable for timber production but support poor lichen crops and therefore do not support mountain caribou (Page, 1985; Stevenson and Hatler, 1985; Simpson and

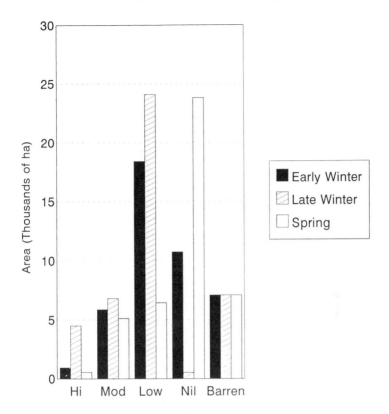

Figure 12.3 Early winter, late winter and spring caribou habitat suitability.

Woods, 1987). Logging roads create access into caribou ranges. An increase in access disrupts normal movement patterns and usually results in an increase in hunting (legal and illegal) and recreational disturbance (Johnson and Todd, 1977; Bloomfield, 1979; Page, 1985; Simpson, 1985; Smith and Herbert, 1985; Stevenson and Hatler, 1985; Simpson and McLellan, 1990).

This GIS-based model has proved valuable to managers as it provides a clear spatial representation of the habitat supposedly utilized by caribou on a seasonal and year-round basis. Since it is GIS-based, the model can easily be modified and redisplayed. It also provides areas of each seasonal range and a good comparison with the total habitat in the study area. While figures on the area available to caribou could be calculated manually in the past, they could not be determined quickly and efficiently, nor could they be updated readily with new knowledge and changes in the forest land base data.

This analysis illustrates the conflict between traditional timber harvest planning and caribou habitat values, as 46 per cent of the planned harvest removes 37 per cent of the high-capability caribou habitat, while 32 per cent of the harvesting planned removes only 5 per cent of the low-capability caribou

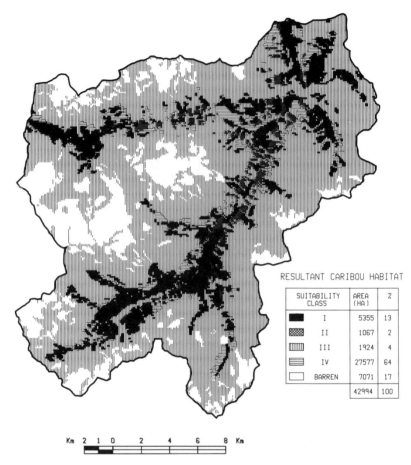

GOLD / BACHELOR WATERSHED

RESULTANT CARIBOU HABITAT

SUITABILITY CLASS		AREA (HA)	%
	I	5355	13
	II	1067	2
	III	1924	4
	IV	27577	64
	BARREN	7071	17
		42994	100

Km 2 1 0   2   4   6   8 Km

Figure 12.4   Critical caribou habitat.

habitat. The availability of operable timber on Class IV caribou habitat provides the opportunity to modify harvesting plans, to assure preservation of
highly suitable caribou habitat and at the same time maintain an adequate
harvesting schedule.

The current caribou habitat model is a first attempt using GIS-based
evaluation methods to assist in the resolution of these forestry/wildlife conflicts.
However, the database and models can be readily improved to provide a
more sophisticated evaluation that can be streamlined to operate on a real-
time basis. The caribou habitat assessment should be expanded to cover
individual and cumulative effects and interactions amongst the major wildlife
resources in the area. To facilitate compatible habitat management for species

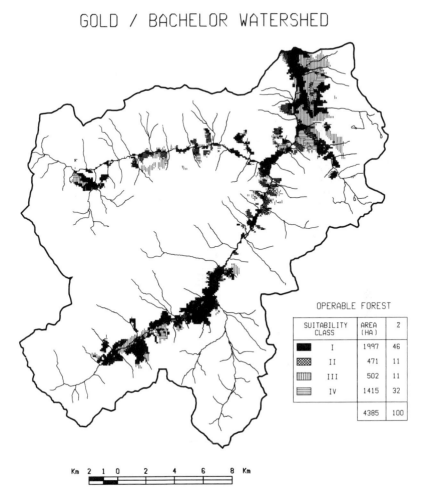

GOLD / BACHELOR WATERSHED

OPERABLE FOREST

SUITABILITY CLASS	AREA (HA)	%
I	1997	46
II	471	11
III	502	11
IV	1415	32
	4385	100

Km 2 1 0 2 4 6 8 Km

Figure 12.5   Operable timber coincident with critical caribou habitat.

other than caribou, potentially limiting habitats need to be identified. It is relatively easy to accommodate the needs of early-successional prolific species such as mule deer and elk into logging plans. Caribou and grizzly bear are sensitive and potentially threatened species within the North Columbia Mountains. Mount Revelstoke and Glacier National Parks are considered one of three possible areas where mountain caribou are likely to persist over the medium to long term in BC, owing to the presence of escape terrain and/or low predator densities. Build-up of prolific early successional species, if accompanied by corresponding increases in predators, is considered a major factor contributing to the ongoing decline of caribou in BC and elsewhere (Seip, 1990). Initially, the suitability of habitat may be analyzed separately

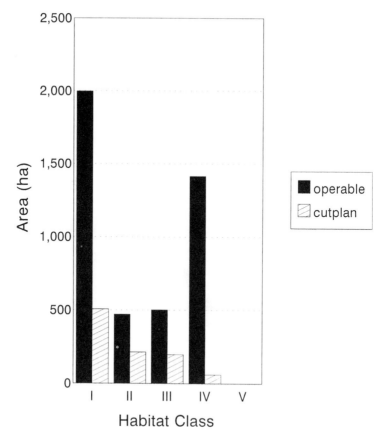

Figure 12.6   Caribou habitat classes (I-IV) coincident with operable forest and the 40-year harvest plan.

for major wildlife species and a combined habitat evaluation could be accomplished using GIS-based overlay techniques. This will identify areas where cumulative shortages of habitat occur during critical seasons. Telemetric data from collared wildlife could be used to verify habitat requirements.

Innovations in harvesting and silviculture prescriptions are needed to address biodiversity. A long-term harvesting model needs to be developed that optimizes regeneration, minimizes fragmentation, maintains diversity, and limits environmental degradation. A system of small, dispersed cut blocks commonly recommended in current guidelines for wildlife, may be incompatible with caribou because it maximizes habitat fragmentation and does not address the old-growth requirements of caribou. The model should consider biophysical and economic components aimed at minimizing disturbance and access, while at the same time maintaining an economically viable annual allowable cutting plan. Such a model is under development (Nelson, 1992), and integration with the wildlife model is planned.

Economic data should be attached to the model so that the costs of different scenarios and trade-offs can readily be determined. This can be accomplished with relative ease for the logging operations, but assigning economic information to wildlife preservation and recreational activities will be challenging. Nevertheless, the current structure of the database and the GIS-based caribou model can be readily adjusted to accommodate some economic evaluation.

Management implications

GIS-based resource quantification provides a foundation for spatial modelling of critical habitats. The spatial and temporal analysis capabilities allow an evaluation of habitat availability, loss, fragmentation and linkage corridors. GIS systems are flexible and permit the evaluation of complex resource issues.

Population goals for a viable caribou herd and the habitat required to sustain such a population need to be defined. Subsequently, harvest scheduling can be modified and trade-off scenarios developed. Forest management should be modified to provide the necessary diversity in habitat over the caribou's home range. Long-term harvesting schedules are required to combine timber and wildlife production with minimum impact on costs and total yields.

Current silviculture practices are aimed at creating dense even-aged forests, which may well lack forage-producing openings and lichen loads sufficient to support caribou. In such forests, recent clearcuts may be the only source of food for species like caribou and moose. Forage on cutovers, however, is likely to be unavailable to caribou and moose during winter, due to heavy snow loads in the Mount Revelstoke and Glacier National Parks area. The implications of providing both snow interception for mobility and openings for forage production in managed forests also need to be assessed and modelled to determine silvicultural and economic implications.

If all resource modelling and assessments are spatially referenced, the GIS can be used as an integrating tool for multi-resource evaluations. Overlay techniques facilitate the identification of conflicts, and provide a new approach for conflict resolution by producing development scenarios for each of the stakeholders, examining the sensitivity of individual variables, and identifying conflicts. Various stakeholders can be accommodated in a neutral system. Such a GIS-based, integrated model is viewed as an ideal tool for decision support when dealing with complex and competing resource uses.

References

Antifeau, T., 1987, The Significance of Snow and Arboreal Lichen in the Winter Ecology of Mountain Caribou (*Rangifer tarandus caribou*) in the North Thompson Watershed of British Columbia. MSc thesis, Simon Fraser University.

Bergerud, A.T., 1978, *The Status and Management of Caribou in British Columbia*, Fish and Wildlife Branch, BC Ministry of Recreation and Conservation.

Bloomfield, M., 1979, Logging/Caribou Relations in the Prince George Area, *in* Dick, J. (Ed.) *Proceedings of a Workshop on Logging/Caribou Relationships with Specific Reference to the North Thompson and Raft PSYUs. Kamloops, May 15–16, 1979*, BC Ministry of Forests and Wildlife Branch, and ELUC Secretariat, BC Ministry of Environment, 25–6.

BC Ministry of Environment and Parks, 1988, *Specifications and Guidelines 1:20 000 Digital Mapping*, Surveys and Resource Mapping Branch.

Fish and Wildlife Branch, 1979, *Preliminary Caribou Management Plan for British Columbia*, Ministry of Environment.

Inventory Branch, 1990, *Standards and Procedures for the Acquisition of Forest Inventory Data*, Resource Inventory Section, BC Ministry of Forests, Victoria.

Johnson, D.R. and Todd, M.C., 1977, Summer use of a Highway Crossing by Mountain Caribou, *Canadian Field-Naturalist*, **91**, 312–14.

Nelson, J.D., 1992, *A Preliminary Analysis of Timber Harvesting Guidelines in the Tangier Drainage*, Forest Operations Research Group, University of British Columbia, 21pp.

Page, R., 1985, Survival Analysis from Radio Telemetry Data, *in* Page, R. (Ed.) *Caribou Research and Management in British Columbia: Proceedings of a Workshop*, Research Branch, BC Ministry of Forests WHR-27 and Wildlife Branch, BC Ministry of Environment RW-41, 241–50.

Seip, D.R., 1990, *Ecology of Woodland Caribou in Wells Gray Provincial Park*, BC Ministry of Environment Wildlife Bulletin No. B-68.

Simpson, K., 1985, The Effects of Snowmobiling on Winter Range Use of Mountain Caribou, *in* Page, R. (Ed.) *Caribou Research and Management in British Columbia: Proceedings of a Workshop*. Research Branch, BC Ministry of Forests WHR-27 and Wildlife Branch, BC Ministry of Environment RW-41, 58–70.

Simpson, K. and McLellan, B.N., 1990, *Wildlife Habitat Inventory and Management Planning in Mount Revelstoke and Glacier National Parks*, Canadian Parks Service.

Simpson, K. and Woods, G.P., 1987, *Movements and Habitats of Caribou in the Mountains of Southern British Columbia*, Wildlife Branch, BC Ministry of Environment and Parks, Nelson, BC Wildlife Bulletin #B-57.

Smith, T.R. and Herbert, D., 1985, Caribou in the Itchas and Ilgachuz: A Summary of Harvest, Inventory and Other Information, *in* Page, R. (Ed.) *Caribou Research and Management in British Columbia: Proceedings of a Workshop*, Research Branch, BC Ministry of Forests WHR-27 and Wildlife Branch, BC Ministry of Environment RW-41, 48–53.

Stevenson, S.K. and Hatler, D.F., 1985, *Woodland Caribou and Their Habitat in Southern and Central British Columbia*, Vol. 1 & 2. BC Ministry of Forests, Land Management Report #23.

Summerfield, B., Escano, R. and Donnelly, B., 1985, *Woodland Caribou Cumulative Effects Analysis Model*, USDA and Idaho Panhandle National Forests.

13

GIS modelling and model validation of deforestation risks in Sagarmatha National Park, Nepal

Gavin Jordan

Introduction

In recent years, there has been debate among researchers in the multi-disciplinary field of Himalayan environmental assessment regarding the causes and severity of Himalayan environmental degradation. Until the mid 1980s, it was widely held that the Himalayan region was experiencing a number of interlinked vicious circles, which were leading to an irrevocable crisis. This has been termed the theory of Himalayan environmental degradation (Ives, 1987). The theory is outlined in the following paragraph; a more detailed account can be found in Ives and Messerli (1989).

Principally because of a reduction in mortality, the population of Nepal has increased rapidly over the past four decades. This has resulted in an increased demand for both forest products, predominantly fuelwood and foliage for fodder, and agricultural land. This has led to a very high rate and degree of deforestation, with some reports suggesting that there will be no accessible forests remaining in Nepal by the year 2000 (World Bank, 1979). Extensive deforestation, particularly on steep hillsides, has led to widespread increases in geomorphic activity, primarily soil erosion through increased overland flow. There are three principal effects: downstream siltation; increases in the magnitude of flooding downstream (i.e. along the Ganges and on its delta) due to increased surface runoff; and removal and/or degradation of the soils of the Nepalese hill farmers, thus reducing productivity per unit area, so that more land has to be put into agriculture. In consequence, more of the existing forest resource has to be cleared, further increasing the distance hill dwellers have to walk to obtain fuelwood, and therefore encouraging the burning of dung, traditionally used as an organic fertilizer. This reduces soil stability and lowers crop yields, encouraging soil erosion, and leads to the conversion of yet more wooded land to agricultural land. By this stage, very steep and infertile areas, which are essentially unsuitable for agriculture, will be utilized for arable cultivation.

This theory of Himalayan environmental degradation has been viewed

as an inexorable process pushing the region toward an increasingly serious
environmental crisis. The theory, widely supported in the scientific literature
of the 1970s and early 1980s, was instrumental in the design and implementation of aid projects for the region. However, in the early 1980s, some of the
assumptions of the theory began to be questioned. Houston (1982, 1987) re-
visited the Khumbu area of Nepal and found the extent of forest cover to be,
at least superficially, as great as on his previous visit three decades earlier. He
also declared that he had photographic evidence to support his statements.
Byers (1987a, b, c) conducted an extensive research programme, partially
based on Houston's early photographs and repeat photography, within the
Sagarmatha National Park. He concluded that the level of environmental
degradation for the area, long regarded as a classic example of tourism-
driven environmental degradation, had been overstated. This, along with other
work, suggested that the theory of Himalayan environmental degradation
was too great a simplification, and that the linkages between individual components of the model were more complex than originally thought (Brower,
1991). It is now felt that the Himalayan region is too heterogeneous for a
simple unifying theory to be applied to it. There is also the serious problem
of a lack of high-quality research data, although this situation is improving in
Nepal, largely due to the Land Resource Mapping Project (LRMP, 1986).

    Given the uncertainty surrounding the seriousness of Himalayan envi-
ronmental degradation, it was decided to conduct a study of the degree of
deforestation risk in the Sagarmatha National Park. This involved the construction of a geographic information system (GIS) model designed to classify
the forest resource into categories reflecting the deforestation risk. The
research had three principal aims:

1. to test the suitability of a GIS for predictive mesoscale modelling of deforestation in a high mountain environment;
2. to ground truth the model, in order to assess how accurate a model could
   be constructed from a remote location, and what aspects of model construction require fieldwork – linked to this, and of significant importance,
   was the gathering of data pertaining to deforestation, i.e. quantitative
   information to assess the level of environmental degradation for this area
   of the Himalaya; and
3. to provide a planning and management tool for the management personnel of the Sagarmatha National Park.

Study area

Sagarmatha National Park is located within the Sagarmatha (Mount Everest)
massif in northeast Nepal, in the Khumbu district on the Tibetan border.
It is home to approximately 6000 Sherpas, who have inhabited the Khumbu
area for several hundred years, since their immigration from Tibet. They and

their animals have altered the vegetation of large areas of the Park through grazing and deforestation. There is a high level of indigenous land management awareness, deriving from the Sherpas' origin on the environmentally-sensitive Tibetan plateau (Stevens, 1993).

The study area is approximately 6 km by 4 km, centred around the Sherpa villages of Namche Bazar, Khunde, and Khumjung. This mountainous area ranges in altitude from 2740 m to 4680 m, with the treeline at approximately 4000 m. The study area consists largely of steep south- and north-facing slopes. At its centre is a plateau on which two of the villages are located. The vegetation cover is principally forest on the north-facing slopes and scrub on the south-facing slopes. This pattern is due partly to site conditions, but largely to a greater degree of human intervention on the more accessible southern slopes. In general, the northern slopes are distant from human habitation, and very steep. Large areas of forests on these slopes show little evidence of human activity or grazing pressure. Much of the forest cover of the plateau is degraded, and the adjacent scrub is used for grazing.

As mentioned above, Sagarmatha National Park has a high level of tourist activity: each year, 7000–8000 visitors come to view some of the highest peaks on Earth. As well as having important social implications (the number of visitors outweighs the number of residents) a great strain is put on the natural resources. Wood is used for the construction of accommodation for tourists and trekkers, heating water for their showers, and for cooking their food. This heavy use of wood is the principal reason why Sagarmatha National Park was used as a prime example by the supporters of the theory of Himalayan environmental degradation.

Model of deforestation

The work conducted for this study included three main stages: the production of a GIS model of deforestation risk undertaken in the UK; ground truthing of the model in Nepal; and analysis of the model, and subsequent remodelling.

Initial model construction

Prior to constructing a model, deforestation risk had to be defined. It was defined as the risk of a loss of forest cover for a particular area, or a significant reduction in the quality of the forest resource, as measured in terms of levels of felling, lopping, and natural regeneration. The percentage canopy cover was measured, to determine if this could be used as an indicator of deforestation risk. Other variables were also included, such as tree height, type and percentage of vegetation cover, and tree species composition.

The model of deforestation risk included four variables, each of which influenced the amount of human utilization that was probable for a particular location. The variables were: proximity to villages; proximity to hamlets; proximity to trails; and slope angle. Each variable directly affects accessibility by people and, to a less significant extent, grazing livestock. Each of the variables formed a different layer of the GIS database, to allow index overlaying (described below) to be performed. In order to enter the four variables into the GIS database, the following spatial information had to be digitized and imported into the GIS (Intera Tydac SPANS version 5.0): contours; trail networks; village locations; and hamlet locations, from a 1:50 000 map of the Khumbu; and vegetation types (forest and non-forest), from a 1:250 000 vegetation map (Dobremez, 1974).

In the initial basic GIS analysis of the imported information, the following were derived: slope angles (from contours); buffer zones around villages; buffer zones around hamlets; and buffer zones along trails.

The buffer zones represented zones of decreasing influence of the village, trail etc. with regard to human activity that would influence the deforestation risk (felling, lopping, fodder gathering, and livestock grazing). The individual layers were then combined to produce the deforestation model, using index overlaying. Index overlaying is a SPANS term for Multiple Criteria Decision Analysis (MCDA), also referred to as Multi-Criteria Evaluation (MCE). This technique is well documented (Carver, 1991; Eastman, 1992; Heywood, 1992). Briefly, index overlaying is a simple modelling technique which allows the combination of a number of map layers, with weightings given to each map, and attribute levels within each map, depending on the perceived importance. For instance, as it was assumed that slope angle and deforestation risk were negatively correlated, gentle slopes were given a higher value than steep slopes. An overlay of vegetation type (forest/non-forest) was included in the index overlay, with non-forest areas being given a negative value. As a negative value in the index overlay procedure precludes that attribute level from the procedure, it was only performed for the forest areas. The index overlay file is in Table 13.1.

At this stage in the model construction, equal weightings were given to the individual layers of deforestation risk. One of the aims of the ground-truthing phase of the study was to assess the relative importance of the individual components influencing deforestation risk, to improve the model.

Index overlaying produced a map with six classes of deforestation risk, which was reclassified to produce three classes to simplify the ground truthing in Nepal. The results of the initial model are shown in Figure 13.1.

Model validation

Ground-truthing of the model was undertaken in Nepal during the post-monsoon season: October–December 1991. Fieldwork was conducted principally to provide sufficient data either to accept the model as an accurate

Table 13.1 Index overlay file for the initial deforestation model

Map 1: Paths (100 and 250 m Buffers) Overlay weighting: 24.0%	*Class*	*Score*
	path	10
	100 m buffer zone	5
	200 m buffer zone	2
	outside buffer zone	0
Map 2: Hamlets and 500 m buffer zones Overlay weighting: 24.0%	*Class*	*Score*
	hamlet	5
	500 m buffer zone	10
	outside buffer zone	0
Map 3: Villages + 2 km buffer zone Overlay weighting: 24.0%	*Class*	*Score*
	village	5
	2 km buffer zone	10
	outside buffer zone	0
Map 4: Slope angle, and consequent erosion risk Overlay weighting: 24.0%	*Slope angle (percentage)*	*Score*
	<10	3
	10–<50	2
	50–<80	1
	>80	0
Map 5: Forest cover Overlay weighting: 4.0%	*Class*	*Score*
	non-forest	−1
	forest	0

The method of calculation for index modelling is

$$\sum_i [(\text{weight of map } i) \times (\text{score for class})] \div (\text{total weight, 100})$$

A negative value excludes a class from index overlay.

representation of deforestation risk in the study area or to allow remodelling to be performed and statistically analysed. A total of 49 sample plots were established. The sampling method used was stratified random sampling. The following variables pertaining to deforestation were recorded at each location: aspect; slope angle; altitude; ground cover: percentage of bare ground, stone, shrub, grass, and leaf litter; tree height; number of seedlings (in 5 m × 5 m plots); lopping (zero, little, moderate or heavy, measured in 10 m × 10 m plots); felling: percentage of trees felled, by stump; canopy cover (percentage).

In addition to these variables, the degree of protection assigned to the sample site by management personnel of Sagarmatha National Park was recorded. The forests of the Park are classified into a number of categories, each permitting specific levels of utilization. The permitted level of utilization ranges from minimal restriction on usage, principally the remote forest areas on the less accessible northern slopes, to a total ban on felling, with severe restrictions on lopping and fodder gathering. Grazing is very difficult to prevent.

As well as quantitative data gathering, a number of interviews were

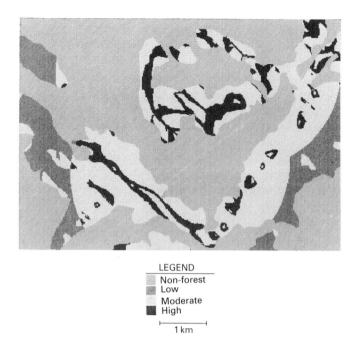

LEGEND
Non-forest
Low
Moderate
High

├─────────┤
1 km

Figure 13.1   Initial deforestation risk model, Sagarmatha National Parks, Nepal.

conducted with Park management personnel, residents, and researchers at the International Centre for Integrated Mountain Development (ICIMOD) in Kathmandu. This provided supportive data and information on both indigenous and imposed management strategies for the Park.

The information gathered in Nepal was analysed and used to test the validity of the model. Correlation coefficients were established for each pair of variables, and scatter plots were produced to find any non-linear relationships (none were found). The deforestation model was statistically supported at the 5 per cent level. Relationships between deforestation risk and percentage felling, and amount of lopping were also significant at this level. There was no significant relationship between deforestation risk and the number of seedlings. Multiple regression analysis determined the strength of relationships between the indicators of deforestation risk, e.g. amount of lopping; per cent felling; canopy cover; and number of seedlings, with the components of the model.

The multiple regression analysis showed that the distance of an area of forest from hamlets and paths was not significantly related to the indicators of deforestation. The strongest relationships were between the indicators of deforestation risk and the level of forest protection, the distance from villages, and the slope angle. However, these had low R^2 values of 0·32, 0·21, and 0·16, respectively, indicating considerable unexplained variation. It was

Table 13.2 Index overlay file for improved deforestation risk model (existing woodlands)

Map 1: Forest protection level	*Class*	*Score*	
Overlay weighting: 46.0%	high	0	
	medium	1	
	low	2	
Map 2: Slope angle, and consequent erosion risk	*Slope angle (percentage)*	*Score*	
Overlay weighting: 23.0%	<10	3	
	10–<50	2	
	50–<80	1	
	>80	0	
Map 3: Villages + 2 km buffer zone	*Class*	*Score*	
Overlay weighting: 31.0%	village	2	
	2 km buffer zone	3	
	outside buffer zone	0	

concluded that a more accurate model could be produced by changing the variables used for model construction and their weightings.

Improved model

A new overlay for the level of forest protection for the forest resource within the study area, information that had been unavailable in the UK, was digitized and index overlaid with maps of slope angle and village location. A new deforestation risk model was developed, with different weightings for the variables. The index overlay weightings are shown in Table 13.2, and the results of the new model in Figure 13.2. The weightings were based on the amount of variability explained by each component of the model; the components explaining the most variability had the greatest weightings. In comparison to the initial model, the new model generally provided stronger relationships between the modelled deforestation risk and the variables indicating deforestation risk (Table 13.3). As can be seen from Table 13.3, considerable unexplained variation remained in the new model. However, it was an improvement on the old model in two respects: it reflected reality to a greater degree, and there were fewer inputs, reducing the time for model construction.

Conclusions

The model

The initial model showed some significant correlations with the level of deforestation risk assessed by fieldwork. These were improved by remodelling.

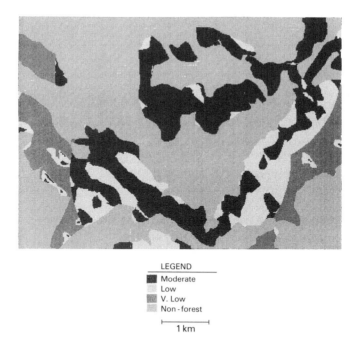

LEGEND
- ▓ Moderate
- ░ Low
- ▒ V. Low
- ▒ Non - forest

⊢————⊣
1 km

Figure 13.2   Improved deforestation risk model, Sagarmatha National Parks, Nepal.

Table 13.3   Statistical relationships between deforestation risk models and indicators of deforestation

	Correlation coefficient	R^2
Amount of lopping		
Initial model	0·374	0·159
Improved model	0·532	0·254
Percent felling		
Initial model	0·374	0·117
Improved model	0·396	0·134
Slope angle		
Initial model	−0·600	0·345
Improved model	−0·575	0·314

Yet there was a large amount of unexplained variation between the models and the deforestation risk assessed from fieldwork. This is believed to be due to the simplistic nature of the model, which does not include important variables such as soil types and rainfall data. The model also does not recognize that deforestation risk is comprised of a number of interconnected components, each of which is influenced by different variables. For example, the amount of felling that occurs in a particular forest stand will be related to a different set of variables than those affecting the number of seedlings per hectare. The amount of felling is due to the direct influence of people, whilst the number of seedlings is principally due to grazing pressure, site conditions, and forest dynamics. This means that a model of the spatial distribution of deforestation risk, as a single dependent variable, is naturally a generalization. A better model of the spatial distribution of deforestation risk could be obtained by modelling the individual components of deforestation risk, and linking them to form an integrated model. Nevertheless, the production of a statistically-significant model of deforestation illustrates the suitability of GIS for this type of work. In a relatively short time, a fairly accurate and useful representation of deforestation risk was produced.

Ground truthing

Ground truthing established that certain variables that were assumed to be salient to deforestation risk were insignificant. Field data allowed an accurate statistical assessment of the model's validity. The period of field research also allowed the collection of information on management strategies, which enabled the construction of a new, more realistic, model and the assessment of its value.

Most GIS models are not validated by fieldwork. Many are created through desk research, and presented and discussed without the collection of any field data or the assessment of the model's accuracy. This is of particular importance in the Himalayas, where the data are notoriously unreliable (Thompson and Warburton, 1985; Byers, 1987a). Both spatial and point data (e.g. meteorological information) can be used to construct an unreliable, yet plausible model. It would have little value, but this could only be determined by ground truthing. If GIS modelling is going to retain and expand its position as an analytical tool for scientific research and environmental management planning, it is vital to assess the accuracy of what is being portrayed.

Level of forest degradation

The initial model assumed that there would be a high level of forest degradation because of the use of wood as a fuel for the indigenous population and visitors, and because of the large amount of recent construction, principally to provide accommodation for visitors. It was also assumed that there would be a scarcity of forest regeneration due to grazing pressure. However, fieldwork showed that the majority of the forests were healthy and, except in certain areas, the deforestation risk was comparatively low. The areas at greatest risk were the central plateau and the zones near the treeline. The plateau is relatively densely populated, and although there are enforced restrictions on forest utilization, there is excessive lopping and felling. Yet, even in these locations, there was a substantial amount of apparently successful natural regeneration. This area of the forest does require a high level of management to ensure that it does not become deforested. Equally, the treeline forests have low growth rates and are particularly vulnerable to deforestation pressures.

Crisis in the Khumbu?

The fieldwork data suggest that certain aspects of the severity of Himalayan environmental degradation have been overstated. It should be stressed that results from one small area of the Himalayas cannot be reliably extrapolated to other areas, as there is great variation in the region. Sagarmatha National Park benefits from a high level of effective and sympathetic environmental management, which is atypical of the majority of high mountain areas of the Himalayas. Paradoxically, areas close to the National Park may have a much greater degree of deforestation risk. Indeed, the restrictions on forestry activities in the National Park probably increase the level of forest utilization in adjacent accessible regions. On the Park's southern limits, there is evidence of extensive harvesting activities, where trees are felled and converted to provide timber for construction within the Park. There are plans to reduce these activities in the future through the creation of a buffer zone, with restrictions on the tolerated level of harvesting, around the Park boundary.

In summary, one can conclude that the level of environmental degradation in Sagarmatha National Park is not as high as was previously thought. This appears to be largely due to successful land management policies of both the Sherpas and the National Park management.

References

Brower, B., 1991, *Sherpa of Khumbu: People, Livestock and Landscape*, New Delhi: Oxford University Press.

Byers, A., 1987a, 'A geomorphic study of man-induced soil-erosion in the Sagarmatha (Mount Everest) National Park, Khumbu, Nepal', unpublished PhD thesis, University of Colorado, Boulder.

Byers, A., 1987b, An assessment of landscape change in the Khumbu region of Nepal using repeat photography, *Mountain Research and Development*, **7**, 77–81.

Byers, A., 1987c, Landscape change and man accelerated soil loss: The case of the Sagarmatha (Mt. Everest) National Park, Khumbu, Nepal, *Mountain Research and Development*, **7**, 209–16.

Carver, S.J., 1991, Integrating multi-criteria evaluation with geographical information systems, *International Journal of Geographical Information Systems*, **5**, 321–39.

Dobremez, J., 1974, *Carte ecologique du Nepal: Kathmandu – Everest 1/250 000*, Paris: Editions du CRNS.

Eastman, J.R., 1992, *IDRISI: A Grid Based Geographic Analysis System*, Version 4.0, Worcester: Clark University.

Heywood, D.I., 1992, '*Multicriteria analysis: Applying expert rules to index overlay*', Proceedings, 1st TYDAC European SPANS Users' Conference.

Houston, C.S., 1982, Return to Everest – a sentimental journey, *Summit*, **28**, 14–17.

Houston, C.S., 1987, Deforestation in Solu Khumbu, *Mountain Research and Development*, **7**, 76.

Ives, J.D., 1987, The theory of Himalayan environmental degradation: Its validity and application challenged by recent research, *Mountain Research and Development*, **7**, 189–99.

Ives, J.D. and Messerli, B., 1989, *The Himalayan dilemma: Reconciling development and conservation*, London: Routledge.

LRMP, 1986, *The Land Resource Mapping Project in Nepal, Agriculture/Forestry report*, Kathmandu and Ottawa: His Majesty's Government of Nepal and the Government of Canada.

Schneider, E., 1978, *Khumbu Himal 1:50 000*, Vienna: Freytag – Berndt und Artaria.

Stevens, S.F., 1993, *Claiming the high ground: Sherpas, subsistence, and environmental change in the highest Himalaya*, Berkeley: University of California Press.

Thompson, M. and Warburton, M., 1985, Uncertainty on a Himalayan scale, *Mountain research and development*, **5**, 115–35.

World Bank, 1979, *Nepal: Development performance and prospects*, Country Study, Washington DC: South Asia Regional Office, World Bank.

PART IV

Simulation and Prediction:
Vegetation and Climate Change

14

Potential impacts of a changing climate on the vegetation cover of Switzerland: a simulation experiment using GIS technology

Bogdan Brzeziecki, Felix Kienast and Otto Wildi

Introduction

Large-scale changes in environmental conditions, such as air pollution and climate change, have caused ecologists to consider the global perspective, and to focus on the heterogeneous distribution of ecological resources, populations, communities, and processes over broad spatial and temporal scales (Binz and Wildi, 1988; Johnson, 1990; Johnston, 1990; Turner, 1990). The study of large regions implies, however, manipulation and analysis of great amounts of geo-referenced data. Recently, it has become possible to handle spatially explicit data of large regions in more efficient, cheap, and user-friendly ways, thanks to technical breakthroughs in the area of Geographical Information Systems (GIS).

Quick access to geo-referenced data and the possibility of generating numerical maps are important but not the only benefits of using GIS in ecology. One interesting and very promising application of GIS technology in environmental studies is to link GIS with static or dynamic simulation models of varying character and complexity. The main prerequisite of such a GIS-assisted modelling effort is the existence of suitable digital data for the area of interest. For Switzerland, fortunately, this requirement is largely fulfilled. Several digitized maps of similar resolution and accuracy, covering the entire country and representing the most important environmental parameters (climate, topography, soil), already exist.

Thus, the objectives of this study are threefold:

1. to create a composite, countrywide environmental database, using GIS software;
2. to develop and interface to the GIS an empirical, probabilistic vegetation-site model, enabling spatial simulation of vegetation patterns;
3. to verify the model and apply it in a simulation experiment on the potential

impact of assumed climate change (raised temperature) on the spatial distribution of forest communities.

Study area

The study area covers the entire Swiss territory with an approximate area of 41 000 km². The climate is mostly temperate-humid, with a strong altitudinal gradient in the Alps. The area can be divided into numerous climatic subregions ranging from intra-alpine dry and continental (total amount of rainfall per year <40 cm) to insubrian (July temperature >20 °C; total rainfall per year >160 cm), or from a temperate lowland climate to a cold high-elevation climate regime above today's timberline. Most vegetated areas at lower elevations have a complex vegetation history and are intensively managed, whereas in the higher zones a more natural vegetation prevails.

Today, approximately 30 per cent of the total area of Switzerland is forested, with a maximum in the Southern Alps (46%) and a minimum in the Central Alps and the Plateau (23 and 24%, respectively). The vegetation zonation is dominated by the altitudinal gradient and is modified by soil and topographic characteristics.

Countrywide landscape-ecological database

The environmental variables currently included in the landscape-ecological database are given in Table 14.1. Data on these variables originate from different sources. The most important database is a digital terrain model (DTM), that provides information on elevation, aspect, and slope for each 250 m × 250 m element of the entire territory of Switzerland. Elevation is the most accurate parameter of the three DTM parameters. Aspect and slope are both derived from elevation, and their accuracy depends on the resolution of the elevation data and the algorithm used to calculate them. A comparison between DTM-values and terrestrial measurements for approximately 11 000 points, established for the Swiss National Forest Inventory, showed a good agreement between measured and calculated DTM elevations. Only 3·1% of points show a deviation between observed and DTM elevations of more than 50 m and only 0·8% show a difference greater than 100 m. However, agreement between actual slope and corresponding DTM values is moderate. As a rule, the number of flat points (slope 0–20%) is overestimated in the DTM, whereas the number of steep slopes (slope > 80%) is underestimated. Nevertheless, the comparison between actual and DTM values shows that for 60% of all points, the deviations do not exceed 20% of inclination; for 74% of the points, 30% of inclination; and only 8% of all points appear to have deviations from measured values greater than 50% of inclination. In the case of aspect, 75% of the points show less than 45° difference between actual and DTM

Table 14.1 The landscape-ecological database; list of site parameters available in a digitized form on a countrywide basis

Site parameter	Data type	Unit	Spatial resolution	Comments
Elevation (DEM)	continuous	m	250 m	
Aspect	continuous	°	250 m	derived from DEM
Slope	continuous	%	250 m	derived from DEM
Mean annual temperature	continuous	°C	1 km²	data interpolated from meteorological stations
Total amount of precipitation	continuous	cm	1 km²	derived from precipitation map 1:400 000
Soil physical parameters:				
– Depth of weathered zone	categorical	1–5	0·5 km²	
– Amount of particles >2 mm	categorical	1–5	0·5 km²	all physical soil parameters derived
– Water capacity	categorical	1–6	0·5 km²	from soil suitability
– Nutrient capacity	categorical	1–6	0·5 km²	map 1:200 000
– Permeability	categorical	1–5	0·5 km²	
– Wetting tendency	categorical	1–4	0·5 km²	
Soil pH	continuous	pH units	1 km²	derived by interpolation from approx. 10 000 Forest Inventory plots

value; in 93% of all points, the DTM values differ no more than 90° from measured values.

To create a complete environmental data set for each DTM point, the elevation data were overlaid with digital climatic maps (temperature and precipitation) and a digital soil suitability map (physical soil properties), with spatial resolutions of approximately 1 km² and 0·5 km² respectively. Soil pH for each DTM point was determined independently by interpolating field measurements from about 11 000 points of the Swiss National Forest Inventory. Since there was no simple spatial interpolation technique available for soil data, a linear interpolation was conducted using routine tools of the ARC/INFO GIS software. Due to this rough method, pH is one of the least accurate variables used. It was retained in the database because of its ecological significance.

Simulated map of the potential natural forest vegetation

Potential natural vegetation

The potential natural vegetation (PNV) of a site can be defined as 'the combination of species that would eventually come into existence under the prevailing environmental conditions of today if man no longer exerted any influence, and if the plant succession had time to reach its final stage' (Ellenberg, 1988; after Tüxen, 1956). Although the present-day plant cover consists, as a rule, of replacement communities that have developed to some degree with the help of man, the PNV is still of general importance as an expression of the natural conditions operating at each site (Ellenberg, 1988). Maps of the potential natural vegetation play an important role in phytosociology, forestry, agriculture, nature and landscape protection, range management, and land use planning. Nowadays, the knowledge of PNV is receiving increasing attention in view of the potential impact of the anticipated climate change on the structure and diversity of plant communities, and on the productivity and ecological potential of habitats.

By definition, the PNV is a hypothetical vegetation, and its interpretation and mapping is often a serious problem (Dierschke, 1974; Bohn, 1981; Ellenberg, 1988; Wildi and Krüsi, 1992). Traditional approaches still underline not only the role of expert knowledge but also of experience in mapping of the potential natural vegetation (Tüxen, 1956; Trautmann, 1966; Kalkhoven and van der Werf, 1988; Falinski, 1990). The weak points of classical vegetation mapping are a low degree of objectivity and reproducibility, as well as high costs (in time and money) because of its labour-intensity. An attempt to overcome some of these weaknesses can be achieved, first, by developing a statistical vegetation-site model, simulating spatial occurrence of vegetation types (Box, 1981; Binz and Wildi, 1988; Davis and Goetz, 1990; Fischer, 1990a; 1990b) and, second, by using GIS for storing, analyzing, and displaying spatially-referenced data.

Vegetation-site model

Under natural conditions, most of Switzerland would be covered by forests. Correspondingly, the dependent variable in the vegetation-site model was defined as the forest community type. The classification used follows the system of forest sites and communities, as suggested by Ellenberg and Klötzli (1972). This hierarchical system includes a total of 71 forest types, defined by means of phytosociological methods. It is widely used in Swiss forestry and silviculture (Leibundgut, 1983), and commonly accepted in mapping forest communities and site conditions (Lienert, 1982; Burnand et al., 1990).

The basic criteria for the selection of the independent variables were, first, the availability of computerized data covering the entire country (with sufficient spatial resolution and accuracy), and, second, the possibility to calibrate the vegetation-site model by means of empirical data (phytosociological relevés). A total of 12 independent variables were included (Table 14.1): two climatic variables (yearly temperature and precipitation), three topographical variables (elevation, aspect, slope), and seven soil variables (soil pH, depth of weathered zone, proportion of soil particles >2 mm, water and nutrient storage capacity, soil permeability, and wetting tendency).

The model that was selected consists of the combination of the parallelepiped (PPD) classifier (Binz and Wildi, 1988) and the Bayes formula (Fischer, 1990a; 1990b). The PPD classifier was applied for discrete variables such as slope, aspect, and soil parameters. For each of these variables, so called attribute matrices, containing 1s and 0s, have been constructed. In the attribute matrix, 0 means that given a specific category of independent variable, a vegetation type is unable to occur, while 1 allows for the occurrence of this type. Thus, for each vegetation unit, an attribute matrix determines qualitatively a possible ecological range (Binz and Wildi, 1988).

In the case of continuous variables (temperature, precipitation, altitude, soil pH), the Bayes formula was used. By means of this formula, the decreasing multivariate state-conditional probability of occurrence was used to order the vegetation types at each point in the geographical space.

To calibrate the model, approximately 7500 phytosociological relevés (data points) were used. Most of these (about 3500 relevés) were obtained from a computerized phytosociological database, covering all major forest types of Switzerland (Sommerhalder *et al.*, 1986; Wohlgemuth, 1992). The main steps of model calibration and development are schematically shown in Figure 14.1; for further details, see Brzeziecki *et al.* (submitted).

Model performance and map evaluation

The vegetation-site model was used to derive vegetation patterns for the entire study area at a resolution of 250 m by 250 m. For the map shown in Plate 12, however, because of its small scale, a 1 km grid was applied. Graphical representation used the GIS software MAPII on Macintosh. While choosing colour scheme, an attempt was made, first, to set off one type of vegetation from another, using as much of the colour spectrum as possible, and second, to choose the colours which can be related to dominating features of the site. Warm colours (yellow, orange, red) are used for communities, that occur on dry sites with a high solar irradiation (*Cephalanthero-Fagion, Quercion pubescenti-petraeae, Molinio-Pinion, Dicrano-Pinion*). Cold colours, e.g. different tints of blue, are applied to communities that occupy excessively wet sites (*Alno-Fraxinion* and related communities, partially also *Piceo-Abietion* and *Vaccinio-Piceion*). Due to the small scale of the map and the large number of vegetation units, the forest communities belonging to the same alliance or suballiance are shown in one colour. As a result of a rather coarse resolution, the map shown in Plate 12 features the general relationships between vegetation, climate and topography. These are thoroughly discussed in Brzeziecki *et al.* (submitted).

To evaluate the performance of the vegetation-site model and to evaluate the quality of the simulated vegetation map, several qualitative and quantitative comparisons have been made between the simulated map and maps prepared using traditional (field) mapping methods.

1. Classification of vegetation: assignment of phytosociological relevés to vegetation types (A,B,C)

grouping

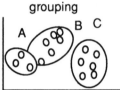

2. Development of the "vegetation-site" model

a) discrete variables

category

	1	2	3	4	5	....
A	1	1	1	0	0	....
B	0	1	1	1	0	....
C	0	1	1	0	0	....

b) continuous variables

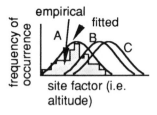

3. Calculation of the probability of occurrence for vegetation types, using country wide digitized maps of site parameters

$$P(V) = f(\text{site factors})$$

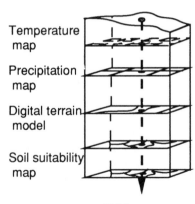

Temperature map

Precipitation map

Digital terrain model

Soil suitability map

4. Importing the simulation results to GIS and constructing the numerical map of the potential natural vegetation

Figure 14.1   Schematic diagram of the major steps performed to obtain the simulated map of the potential natural forest vegetation of Switzerland.

Qualitative assessment of the simulation results has involved the comparison of the simulated occurrence of particular vegetation types with empirical maps, showing the spatial distribution of corresponding phyto-sociological relevés. Figure 14.2 shows the results of such a test for two communities with contrasting ecological amplitudes and altitudinal ranges: *Galio odorati-Fagetum typicum* (colline/submontane) and *Larici-Pinetum cembrae* (subalpine). It can be noted that at the coarse scale considered, there is a rather good correspondence between empirical data and simulated patterns.

For the quantitative comparison, the 1:50 000 map representing the spatial distribution of natural forest communities in the Canton of Obwalden (Lienert, 1982), was selected. The map appeared to be well-suited for a comparison test for the following reasons: the scale and thematic resolution of the map match those of the simulated map; the forest types of the region cover a wide range of vegetational variation, as a result of the large altitudinal gradient and diverse topographic and soil conditions; and the vegetation classification that is used in the map is basically the same as applied in the model.

The comparison between simulated and charted vegetation types was made on the basis of 524 sample points, that were selected at a distance of 250 m along north-south transects of 1 km length. The test is conducted at the level of forest community (Table 14.2). Only climax and semi-climax communities were taken into account; the communities that occupy extreme environments and, as a rule, seldom occur on the map area, were excluded. The degree of correspondence varies between community types. On average, it amounts to 45.8 per cent if the community with the highest probability of occurrence is considered. Taking into account the second and third probability of occurrence, the model simulates the true community in an additional 20 per cent and 12 per cent of all cases, respectively. Thus, there is a chance of about 80 per cent that, within the first three communities, the appropriate one is simulated.

Climate change experiment

The potential natural vegetation of a site is determined by the entire set of ecological factors operating at this site, and therefore it can be considered an ideal parameter expressing the ecological potential of habitat. Thus, one of the most interesting applications of the numerical map of potential natural vegetation is to use it in simulation experiments, aimed at determining the potential impact of environmental change on the ecological potential of forest sites. The results of such a modelling experiment, involving a general temperature change of 2 °C (annual mean temperature) are summarized in Figures 14.3–14.5 and in Table 14.3. To simplify the analysis, only two major groups of communities, differing significantly according to their species

Galio odorati–Fagetum typicum

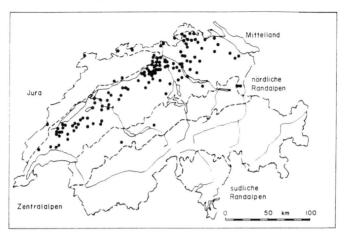

(a)

Galio odorati–Fagetum typicum

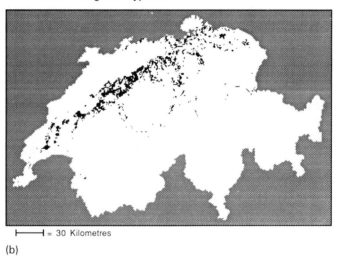

⊢——⊣ = 30 Kilometres

(b)

Figure 14.2   Comparison between empirical and simulated geographical distribution of Galio-Fagetum *vegetation type. a) Empirical spatial distribution of phytosociological relevés, according to Ellenberg and Klötzli (1972). b) Simulated spatial distribution; black dots indicate pixels where a given vegetation type occurs with the highest probability.*

Larici–Pinetum cembrae

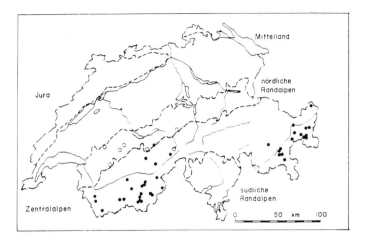

(a)

Larici–Pinetum cembrae

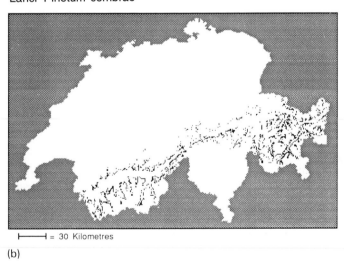

⊢———⊣ = 30 Kilometres

(b)

Figure 14.3 Comparison between empirical and simulated geographical distribution of Larici-Pinetum *vegetation type. a) Empirical spatial distribution of phytosociological relevés, according to Ellenberg and Klötzli (1972). b) Simulated spatial distribution; black dots indicate pixels where a given vegetation type occurs with the highest probability.*

Table 14.2   Comparison between field mapping and the simulated vegetation map for the Canton Obwalden

Forest community type	Simulation priority					
	n	1	2	3	Σ	%
Galio-Fagetum typicum	11	3	3	3	9	81·8
Milio-Fagetum	23	10	4	3	17	73·9
Pulmonario-Fagetum typicum	27	8	6	2	16	59·2
Cardamino-Fagetum typicum	59	19	21	5	45	76·3
Carici-Fagetum typicum	5	2	1	1	4	80·0
Abieti-Fagetum typicum	119	57	29	21	107	89·9
Abieti-Fagetum polystichetosum	16	10	2	1	13	81·2
Aceri-Fagetum	2		1		1	50·0
Bazzanio-Abietetum	15	5	2	1	8	53·3
Equiseto-Abietetum	30	6	3	4	13	43·3
Adenostylo-Abietetum	51	21	15	6	42	82·4
Sphagno-Piceetum calamagrost.	133	93	13	11	117	88·0
Piceo-Adenostyletum	33	6	7	5	18	54·5
Σ	524	240	107	63	410	
%		45·8	20·4	12·0	78·2	

Only climax and semi-climax forest communities are included; *n*: number of sampling points; Σ and %: absolute and relative number of geographical points where the simulated community (1, 2, or 3 priority) is identical with the community on the field map.

composition, environmental conditions, productivity and silvicultural treatment, are considered.

The first group, including 21 forest communities, is distinguished by a dominance of beech (*Fagus sylvatica*). The ecological amplitude of beech-dominated communities in Switzerland is rather broad and ranges from the colline to the subalpine belt. Beech-dominated communities constitute, as a rule, zonal vegetation. They occur under various soil and topographical conditions with the exception of the most extreme sites (e.g. excessively wet soils, very shallow and stony soils, steep and/or dry slopes).

The second group includes 12 communities characterized by the dominance of spruce (*Picea abies*). Under natural conditions, the distribution of zonal communities of spruce-dominated forests is largely limited to the subalpine and, partially, to the montane vegetation belt (the latter in the Central Alps). The simulated spatial distribution of these two groups, under today's climate and soil conditions, is shown in Figure 14.4.

In the experimental run of the vegetation-site model, at each pixel point, a temperature increase of 2 °C was assumed. The results of this experimental simulation were compared with those obtained for today's conditions (control run). The spatial arrangement of pixels, distinguished for the shift from one community group to another between control and experimental model

Fagus sylvatica

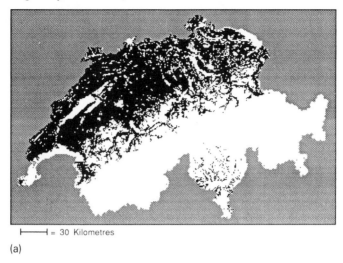

⊢——⊣ = 30 Kilometres

(a)

Picea abies

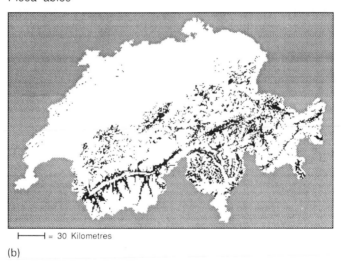

⊢——⊣ = 30 Kilometres

(b)

Figure 14.4 Simulated spatial distribution of communities dominated by a) Fagus sylvatica *or b)* Picea abies *under today's climatic conditions.*

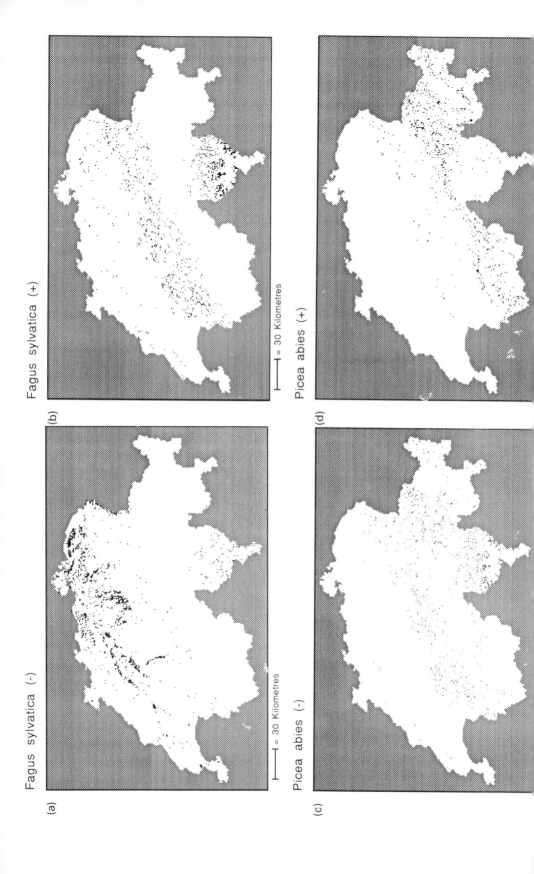

Fagus sylvatica (−)

Fagus sylvatica (+)

⊢——⊣ = 30 Kilometres

(a)

(b)

Picea abies (−)

Picea abies (+)

⊢——⊣ = 30 Kilometres

(c)

(d)

Table 14.3 Numbers and percentages of pixels, where either vegetation types dominated by Beech (Fagus sylvatica) or Spruce (Picea abies) are simulated, under 'today's' and 'increased temperature' conditions

Pixel category	'Today's' climate	Temperature increase (+2 °C)	
		Beech community	Non-beech community
Beech community	16767 56·4%	15305 51·5%	1462 4·9%
Non-beech community	12940 43·6%	1067 3·6%	11873 40·0%
Total	29707 100%	16372 55·1%	13335 44·9%

Pixel category	'Today's' climate	Temperature increase (+2 °C)	
		Spruce community	Non-spruce community
Spruce community	5995 20·2%	4727 15·9%	1268 4·3%
Non-spruce community	23712 79·8%	816 2·7%	22896 77·1%
Total	29707 100%	5543 18·6%	24164 81·4%

runs, is shown in Figure 14.5 (only shifts from beech to non-beech and from spruce to non-spruce communities are considered). It can be seen that the pixels characterized by a loss of the beech communities tend to accumulate in the two regions: the Plateau (especially the northeastern part) and the Southern Alps. On the contrary, the pixels distinguished by a loss of spruce communities are predominantly located in the Northern Prealps, as well as in Southern and Central Alps. The spatial distribution of pixels characterized by a gain of beech communities agrees largely with the distribution of pixels distinguished by a loss of spruce communities. The pixels distinguished for a gain of spruce communities are to a great extent limited to the Central Alps.

The altitudinal distributions of the loss and gain pixels reveal different patterns for beech- and for spruce-dominated communities (Figure 14.6). In the case of beech communities, the loss and gain pixels occupy two distinctive altitudinal ranges. The loss pixels lie predominantly in the narrow altitudinal range of 500–700 m. Under today's climate, beech-dominated forests find here, as a rule, excellent growth conditions. As a result of the increased temperature, however, beech and some of its accompanying species may lose competitive

Figure 14.5 Spatial distribution of pixels, that show either a loss (–) or a gain (+) of the Fagus sylvatica- or Picea abies-dominated communities in a climate change experiment.

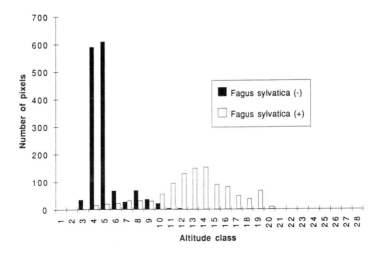

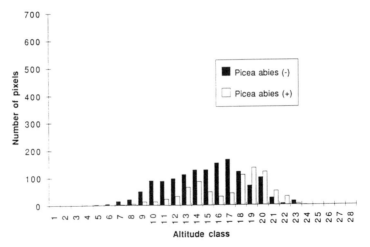

Figure 14.6   Altitudinal distribution of pixels that show either a loss (–) or a gain (+) of the Fagus sylvatica- *or* Picea abies-*dominated communities, in a climate change experiment. Altitude classes are: 1:200–300 m; 2:300–400 m.*

ability in comparison to the more thermophilous and drought-resistant species of broad-leaved mixed woods (especially from the alliances *Carpinion* and *Quercion pubescenti-petraeae*). In contrast to the loss pixels, the gain pixels occupy a rather wide altitudinal range (most of them lie between 1100 and 2100 m). Under today's temperature regime, these elevations are generally too cold for the growth of beech. However, with an increased temperature and less severe climate, beech may become more competitive in these locations.

   In the case of spruce forests, the ranges of loss and gain pixels are similar and are not as well separated as for beech communities. Probably, it is the result of the broad ecological amplitude of spruce communities, and of considering different geographical regions of Switzerland simultaneously, characterized by various climatic regimes. This pattern could possibly be changed by analysis at the regional level. Nevertheless, the results presented in Figure 14.6 suggest that there is a similar tendency as observed with the beech communities, i.e. spruce woods become less competitive at lower, and more competitive at higher elevations.

Numbers of pixels belonging to the categories described above are summarized in Table 14.3. It can be noted that for both beech and spruce communities, the total numbers of pixels remain rather stable, i.e. the numbers of losses and gains are similar (for beech-dominated communities, 1462 losses versus 1067 gains, and for spruce-dominated communities, 1268 losses versus 816 gains). For beech communities, the net loss amounts to 395 pixels, or 2.4 per cent of beech-dominated pixels under today's conditions, and for spruce communities – 552 pixels, or 9.2 per cent of spruce-dominated pixels). This would suggest that significantly higher net gains or losses can only be expected for those communities which occupy more extreme positions on the climate and elevational gradient than the two groups of communities considered here.

Discussion and conclusions

Until some years ago, the only way to store and present the spatial vegetational patterns was the classical map: a two-dimensional piece of paper with colours and/or symbols (Van der Zee and Huizing, 1988). Since the breakthrough in the field of GIS in the late 1980s, it has become possible to store such data in a computerized form, ready for immediate retrieval in map form (Van der Zee and Huizing, 1988; Johnson, 1990). The vegetation data can be entered into a GIS in different ways. One of the most interesting approaches is to develop a vegetation-site model, simulating the occurrence of vegetation types as a function of different site factors, and to interface it with a GIS. Recently, several vegetation-site models have been developed and applied to simulate the distribution of different kinds of vegetation, at different spatial scales (Box, 1981; McCune and Allen, 1985; Binz and Wildi, 1988; Davis and Goetz, 1990; Fischer, 1990a; 1990b). In the present study, the combination of methods proposed by Binz and Wildi (1988) and Fischer (1990a, 1990b) was used.

The disagreements between simulated and mapped vegetational patterns which appeared in the course of the testing procedure, may be attributed to several factors, including errors associated with field maps and incomplete and low-representative empirical data used to calibrate model parameters. These problems are discussed in more detail in Brzeziecki *et al.* (submitted). At the moment, the most serious limitation for obtaining a higher-quality vegetation map appears to be the low resolution and insufficient accuracy of the input (environmental) data. The variable with the best resolution (250 m) and the greatest accuracy is elevation. The accuracy and resolution of all other factors (including slope and aspect) are significantly lower. In the case of climatic factors (temperature and precipitation), the relatively low resolution seems to be a minor problem, since both parameters vary continuously in the geographical space. In the case of soil parameters which are characterized by abrupt changes in the geographical space, the low data resolution is the main obstacle in obtaining a better fit between simulated and mapped types. Moreover, because of the small scale of the soil map (1:200 000), the mapped units aggregate soil types with different properties. Thus, the differences within the zonal vegetation groups, which are mainly soil-dependent, would require more detailed input information to be explicitly simulated. More detailed digitized maps of soil parameters (e.g. maps of soil types) at a scale of 1:25 000 to 1:50 000 would be adequate for the resolution of our model.

In spite of these difficulties, the results of the simulation seem to meet well the level of precision required to perform risk assessment studies of climate change, or to provide forest management guidelines on the regional level. As shown above, prediction of the geographical ranges of plant communities under expected climatic change, can be based on correlations between the distribution of communities and environmental factors, as determined under today's conditions. It should be stressed, however, that this static regression-like approach is only valid within the range of the empirical data used to calibrate the model. Consequently, the interpretation of individual pixels whose values lie outside the calibration range has to be done with care.

In general, the results of the temperature simulation experiment seem to be plausible. Due to the strong altitudinal and climatic gradient, a temperature increase must lead to a shift of the zonal (climate-dependent) communities from lower to higher elevations. Developing the vegetation-site model and linking it to GIS enables us to determine various quantitative characteristics and spatial aspects of the potential vegetational response.

From this point of view, the numerical map of potential natural vegetation has a big advantage over the classical map. The latter, once prepared, is a final document that cannot be changed or converted after it has been printed. On the contrary, the simulated map has a dynamic and functional character, and can easily be used as a tool for simulation experiments. The interpretation and practical value of quantitative predictions are, however, still dependent on the quality of the ecological model used, as well as on the resolution and accuracy of the input data of environmental parameters.

References

Binz, H.R. and Wildi, O., 1988, Das Simulationsmodell MaB-Davos. Schlussbericht zum schweizerisches MaB-Programm Nr. 33. Bundesamt für Umweltschutz, Bern.
Bohn, U., 1981, Vegetationskarte der Bundesrepublik Deutschland 1:200 000 – Potentielle natürliche Vegetation – Blatt CC 5518 Fulda, Schriftenr. Vegetationskunde (Bonn-Bad Godesberg) 15: 1–330.

Box, E.O., 1981, *Macroclimate and Plant Forms: An Introduction to Predictive Modelling in Phytogeography*, The Hague: Junk.

Brzeziecki, B., Kienast, F. and Wildi, O. (submitted), Simulated map of the potential natural vegetation of Switzerland, *Journal of Vegetation Science*.

Burnand, J., Hasspacher, B. and Stocker, R., 1990, *Waldgesellschaften und Waldstandorte im Kanton Basel-Landschaft*, Verlag Kanton Basel, Liestal.

Davis, F.W. and Goetz, S., 1990, Modeling vegetation pattern using digital terrain data, *Landscape Ecology*, **4**, 69–80.

Dierschke, H., 1974, Zur Abgrenzung von Einheiten der heutigen potentiell natürlichen Vegetation in Waldarmen Gebieten Nordwestdeutschlands, in Tüxen, R. (Ed.) *Tatsachen und Probleme der Grenzen in der Vegetation*, Verlag J. Cramer, Lehre, 305–25.

Ellenberg, H., 1988, *Vegetation Ecology of Central Europe*, Cambridge: University Press.

Ellenberg, H. and Klötzli, F., 1972, Waldgesellschaften und Waldstandorte der Schweiz, *Mitteilungen der schweizerischer Anstalt für das forstliche Versuchswesen*, **48**, 388–930.

Falinski, J.B., 1990, Kartografia geobotaniczna, Panstwowe Przedsiebiorstwo Wydawnictw Kartograficznych, Warszawa-Wroclaw.

Fischer, H.S., 1990a, Simulating the distribution of plant communities in an alpine landscape, *Coenoses*, **5**, 37–43.

Fischer, H.S., 1990b, Simulation der räumlichen Verteilung von Pflanzengesellschaften auf der Basis von Standortskarten. Dargestellt am Beispiel des MaB Testgebietes Davos, Dissertation ETH Nr. 9202.

Johnson, L.B., 1990, Analyzing spatial and temporal phenomena using geographical information systems, *Landscape Ecology*, **4**, 31–43.

Johnston, C., 1990, GIS: more than a pretty face, *Landscape Ecology*, **4**, 3–4.

Kalkhoven, J.T.R. and van der Werf, S., 1988, Mapping the potential natural vegetation, in Küchler, A.W. and Zonneveld, I.S. (Eds) *Vegetation mapping*, Dordrecht: Kluwer, 375–86.

Leibundgut, H., 1983, Die waldbauliche Behandlung wichtiger Waldgesellschaften der Schweiz, *Mitteilungen der schweizerischer Anstalt für das forstliche Versuchswesen*, **59**, 1–78.

Lienert, L. (Ed.) 1982, *Die Pflanzenwelt in Obwalden*. Ökologie, Kant. Oberforstamt Obwalden, Sarnen.

McCune, B. and Allen, T.F.H., 1985, Will similar forests develop on similar sites? *Canadian Journal of Botany*, **63**, 367–76.

Sommerhalder, R., Kuhn, N., Biland, H.-P., von Gunten, U. and Weidmann, D., 1986, Eine vegetationskundliche Datenbank der Schweiz, *Botanica Helvetica*, **96**, 77–93.

Trautmann, W., 1966, Erläuterungen zur Karte der potentiellen natürlichen Vegetation der Bundesrepublik Deutschland, Blatt 85 Minden, Schriftenr. *Vegetationskunde (Hiltrup i.W.)*, **1**, 1–138.

Turner, M.G., 1990, Spatial and temporal analysis of landscape patterns, *Landscape Ecology*, **4**, 21–30.

Tüxen, R., 1956, Die heutige potentielle natürliche Vegetation als Gegenstand der Vegetationskartierung, *Angewandte Pflanzensoziologie (Stolzenau/Weser)*, **13**, 5–42.

Van der Zee, D. and Huizing, H., 1988, Automated Cartography and Electronic Geographic Information Systems, in Küchler, A.W. and Zonneveld, I.S. (Eds) *Vegetation mapping*, Dordrecht: Kluwer, 163–89.

Wildi, O. and Krüsi, B., 1992, Revision der Waldgesellschaften der Schweiz: Wunsch oder Notwendigkeit? *Schweizerische Zeitschrift für Forstwesen*, **143**, 37–47.

Wohlgemuth, Th., 1992, Die vegetationskundliche Datenbank, *Schweizerische Zeitschrift für Forstwesen*, **143**, 22–36.

15

GIS analysis of the potential impacts of climate change on mountain ecosystems and protected areas

Patrick N. Halpin

Introduction

The evaluation of the potential impacts of future climatic changes on mountain ecosystems and nature reserves has become increasingly important in the study of the long-term management and protection of biodiversity (McNeely, 1990; Peters and Darling, 1985). The possibility of rapid changes in the climatic habitats of mountain ecosystems due to emissions of greenhouse (radiatively-active) gases raises many difficult questions concerning the vulnerability of mountain nature reserves to environmental disruption in the future. Proposed changes in global temperatures and local precipitation patterns could significantly alter the altitudinal ranges of important species within existing mountain nature reserves and create additional environmental stresses on already fragile ecosystems.

Until recently, generalized speculations concerning the possible future movement of ecoclimatic habitats within mountain nature reserves have been made without reference to actual mountain sites or the application of geographic modelling methods (Peters and Darling, 1985; Hunter *et al.*, 1988; Graham, 1988). However, current analysis techniques using geographic information systems (GIS) now allow the simulation of possible responses of existing mountain nature reserve systems to both global and regional climate change (Halpin, 1992; Leemans and Halpin, 1992; Halpin, 1993). This type of geographic analysis provides for the development and testing of more detailed hypotheses concerning global change and mountain environments.

This brief overview paper outlines a GIS pilot study which assesses the potential impacts of various future climate change scenarios on the global distribution of mountain vegetation regions, the possible impacts on important mountain nature reserves, and a regional case study of impacts on a tropical mountain area. It also presents a hypothetical analysis of the differences in the sensitivity of mountain reserve areas to climatic change due to their latitudinal position. This pilot study emphasizes the role of GIS modelling techniques in the analysis of this problem, and suggests possible methods of analysis for future impact studies.

Conservation in changing climates: past approaches

The potential impacts of rapid climatic change on natural ecosystems and nature reserves strongly illustrate the need for the development of creative research techniques for the analysis of the potential movement of habitats within the confines of existing nature reserve boundaries. In the most basic form, rapid changes in climatic conditions coupled with ongoing changes in landscape fragmentation are seen to be two potentially interactive environmental forces. These combined forces could drastically hinder the ability of natural resource managers to maintain viable mountain habitats and species populations in the future.

Peters and Darling (1985) outlined two basic hypotheses concerning potential climatic changes and nature reserves. The first general hypothesis defines the interaction of landscape fragmentation and species migration under changing climates. Figure 15.1 depicts this interaction by illustrating the present climatic range of an endangered species in relationship to a proposed reserve area, the same reserve area after habitat fragmentation, and the isolated reserve after a shift in the climatic range of the species. The basic premise of this general hypothesis is that under changing land use and climatic conditions, static nature reserve boundaries may fail to encompass the species ranges they were intended to protect. The explicit process outlined in this hypothesis is that the southern climatic ranges of species in the northern hemisphere would move poleward under proposed climate change scenarios in direct response to increasing temperatures. The ability of species to track changing environmental conditions would be related to the rate of climatic change, the migratory potential of the species, competitive pressures between species, and physical obstatcles in the path of migrating individuals (Smith and Tirpak, 1989; Solomon, 1986; Davis, 1989; Sedjo and Solomon, 1989; Hunter *et al.*, 1988; Graham, 1988).

The need to identify potential boundaries to the migration of species between reserve areas has spurred a wide interest in the geographical (GIS) analysis of the distributions and connectiveness of protected areas in recent years (Mackintosh *et al.*, 1989; Scott *et al.*, 1990; Hudson, 1991; Shafer, 1992). This type of analysis has led to the development of a variety of rule-of-thumb assumptions concerning the spatial connectiveness of nature reserves under changing climatic conditions. Noss (1992) states that corridors between reserves should be aligned up-slope, coast-inland, and south-north in order to facilitate potential species movements under changing climatic conditions. This type of generalization has been coupled with other basic biogeographic theories to form the basis for numerous landscape-level GIS projects now investigating the effects of fragmentation on natural resource management (Scott *et al.*, 1990; Hudson, 1992).

The second general hypothesis presented by Peters and Darling (1985) and others concerns the potential movement of the climatic ranges of species along altitudinal gradients. Under this general hypothesis, increases in local

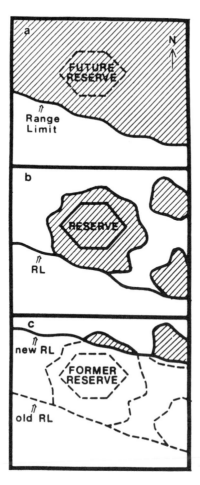

Figure 15.1 A conceptual model of the potential impacts of landscape fragmentation and climatic change on species ranges and nature reserve boundaries (adapted from Peters and Darling, 1985).

temperature would act to move the climatic habitats of species up-slope, in a linear manner, where each successive altitudinal range would be replaced by the species habitat occupying the zone directly below (Figure 15.2). This conceptual model is based on the general premise that the boundaries of species' present climatic ranges will respond symmetrically to changes in temperature related to the adiabatic lapse rate (the loss of temperature with increasing altitude) for a particular mountain site. The general biogeographical rule used to derive this conceptual model is attributed to Hopkins bioclimatic law (Peters and Darling, 1985; MacArthur, 1972), which relates a 3 °C change in temperature to a 500 m change in altitude. Under this general conceptual model, the expected impacts of climate change in mountainous nature reserves

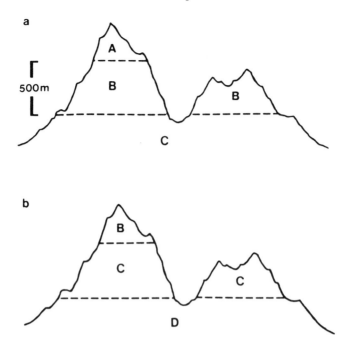

Figure 15.2 A conceptual model of the potential movement of species climatic ranges along altitudinal gradients under a climatic change scenario. Movement of climatic ranges is based on a 500 m increase in altitude for each 3 °C increase in temperature (adapted from Peters and Darling, 1985).

would include the loss of the coolest climatic habitats at the peaks of the mountains and the linear migration of all remaining habitats up-slope.

Peters and Darling's (1985) conceptual model was not intended to represent an operational model of altitudinal change for all mountain sites. Rather, it was presented as a heuristic representation of a possible outcome of climate change that emphasized risks to upper mountain zones (Peters, personal communication). However, these extremely generalized assumptions have been incorporated directly into much of the current literature on potential impacts of climatic change on the management of biodiversity and nature reserve systems (Smith and Tirpak, 1989; McNeely *et al.*, 1990; IPCC, 1989; Noss, 1992). The general management conclusion derived from this theory of species response is that nature reserves exhibiting large altitudinal ranges and topographic relief will offer the largest range of climatic habitats, and therefore will allow for the greatest amount of internal species movement under changing climates. In contrast, climatic shifts in areas of low altitudinal range would be expected to force species to migrate outside the boundaries of the reserve area.

This conceptual model has been used as the underpinning for the analysis of the altitudinal ranges of global nature reserve networks in an initial

Table 15.1   Altitudinal range of protected areas (from: McNeely, 1990)

Biogeographic Realm	Altitudinal Range (in meters)							Totals
	0 999	1000 1999	2000 2999	3000 3999	4000 4999	5000 5999	6000 6999	
Nearctic	171	41	27	4	6	2	2	253
Palearctic	423	146	49	25	6	4	4	657
Afrotropical	319	50	14	2	2	–	–	387
Indomalaya	346	92	21	7	–	–	–	466
Oceania	23	8	1	2	2	1	–	37
Australia	85	26	–	–	–	–	–	111
Antarctic	76	21	4	3	–	–	–	104
Neotropical	161	51	34	21	5	3	–	277
Totals	1604	435	150	64	21	10	6	2290

assessment of the risks posed to these reserves by changing climatic conditions. McNeely (1990) assessed the altitudinal ranges of more than 2000 nature reserves greater than 1000 hectares in size (Table 15.1), determining that 686 (approximately 30 per cent) of the sites assessed had an altitudinal range of more than 1000 m. Using the rule of thumb of 3 °C temperature change per 500 m of altitude, these reserves are then assumed to contain approximately 6 °C of potential climatic range. This temperature range is assumed to be within the range of most future climate change scenarios. To quote McNeely (1990): such areas appear to contain sufficient altitudinal variation to accommodate the climate changes forseen in the coming century.

This direct extrapolation of intuitive theories of expected biogeographic response to management-oriented assessments, without geographic analysis of the spatial characteristics of potential climate changes on mountainous terrain, needs more rigorous evaluation. While it is generally accepted that larger altitudinal gradients potentially offer a larger number range of ecoclimatic habitats, the implicit assumption that changes in the climate will offer a similar number and spatial configuration of distinct climatic habitats along an altitudinal gradient needs reconsideration.

GIS approaches

This GIS pilot study assesses the global distribution of potential climatic impacts for a wide range of nature reserves and altitudinal ranges outlined above. It will show that the general assumptions used in previous analyses do not correspond to currently accepted biogeographical principles and that the simplistic rules of thumb used in past assessments may result in misleading conclusions (Halpin, 1992, 1993).

First, all climate change scenarios are not spatially equivalent, so that assessments of the expected impacts of projected climatic changes on mountain nature reserves must be assessed in terms of their location with respect to variation in the magnitude of the climate changes at specific locations. Second, all ecoclimatic habitats are not equivalent. For example, the potential climatic range or extent of a tropical montane forest and a northern alpine larch forest are significantly different, and any conceptual model which treats the responses of different climatic habitats as symmetrical is destined to misrepresent the composition of altitudinal change for different latitudes and biogeographic regions. Third, the actual complexity of population level dynamics which would occur as individual species attempt to track shifting climates onto infertile, high-angle alpine areas occupied by other persistent species is an extremely difficult problem to address with simple bioclimatic rules.

Global methodology

As a first step in a GIS analysis of potential impacts of climate change on mountain ecosystems and nature reserves, a global grid model was used to establish current mean climatic conditions for the world. The $0.5° \times 0.5°$ climatic grid GIS developed by Leemans and Cramer (1990), which contains information on monthly average temperature, precipitation, cloudiness, soil texture, soil water holding capacity, and elevation, was used to develop an ecoclimatic classification (potential vegetation) for the world (Prentice *et al.*, 1989). The climatic grid GIS was then used to derive a Holdridge life zone class and an effective soil moisture budget for each grid cell.

The Holdridge life zone model is a widely-used global ecoclimatic classification system based on the relationship of current vegetation biomes to three general climatic parameters: annual temperature, annual precipitation and an estimated potential evapotranspiration presented on logrithmic axes (Holdridge, 1967). The model predicts ecoclimatic areas currently associated with major vegetation formations of the world, and does not directly model actual vegetation or land cover distribution.

In order to test the potential impacts of changing climatic conditions at a global scale, the differences in monthly temperature and precipitation for four General Circulation Models (GCMs) were geometrically interpolated and overlaid into a raster GIS system to develop data layers of new distributions of equilibrium ecoclimatic zones for each scenario (Emmanuel *et al.*, 1985; Smith *et al.*, 1990; Smith *et al.*, 1992). The scenarios evaluated present a wide range of different climate change modelling experiments from different research institutions. Table 15.2 presents the general features of each of these climatic change scenarios.

While all of the climate change scenarios tested present different spatial patterns and magnitudes of change, a few general trends were observed. The highest percentage of change occurred in the high northern latitudes under

Table 15.2   General circulation models used to construct climate change scenarios

GCM	Change in mean global:	
	Temperature (°C)	Precipitation (%)
Oregon State University (OSU)[a]	2.84	7.8
Geophysical Fluid Dynamics Laboratory (GFDL)[b]	4.00	8.7
Goddard Institute for Space Studies (GISS)[c]	4.20	11.0
United Kingdom Meteorological Office (UKMO)[d]	5.20	15.0

[a] Schlesinger and Zhao, 1988.
[b] Manabe and Wetherald, 1987.
[c] Hansen *et al.*, 1988.
[d] Mitchell, 1983.

all scenarios, due to higher temperature forcing at these latitudes (Kalkstien, 1991). This feature becomes important when analysing potential impacts on existing nature reserves, which have a numerical bias toward more northerly, developed countries (Halpin, 1992; Smith *et al.*, 1990). To illustrate this point, the locations of 299 Biosphere Reserves were digitized into the global GIS system, assigned current climate ecoclimatic classes, and compared to the ecoclimatic classes of the sites for each change scenario. A change in full Holdridge ecoclimatic class was assumed to represent a significant impact for each site (Halpin, 1992; Smith *et al.*, 1990; Leemans and Halpin, 1992). This process was repeated for a larger database of 2290 reserves listed by the World Conservation Monitoring Center (WCMC) which provided similar results (Leemans and Halpin, 1992). Under both sets of analysis, the distribution of impacts on nature reserves was higher than for terrestrial biomes in general due to the uneven distribution of reserve sites (Halpin, 1992).

A number of nature reserves with an altitudinal range of more than 3000 m were plotted and compared to the spatial distribution of climatic change under the four scenarios tested (Figure 15.3). Table 15.3 presents the percentage of these reserves which experience a change in ecoclimatic zone under each climate change scenario. As with the general distribution of important nature reserves, high mountain reserves received significant levels of impacts under all scenarios tested.

In order to assess the spatial distribution of climate change scenarios against global mountain ranges, a GIS data layer of mountain vegetation regions of the world was digitized and overlaid onto the ecoclimatic change scenario data layers (Figure 15.4). As with the general distribution of nature reserves, global mountain systems displayed a higher percentage of areal ecoclimatic change than terrestrial ecosystems in general. Table 15.4 depicts the percentage of area in mountain and non-mountain environments changing ecoclimatic zone. These elevated rates of ecoclimatic impact can be attributed

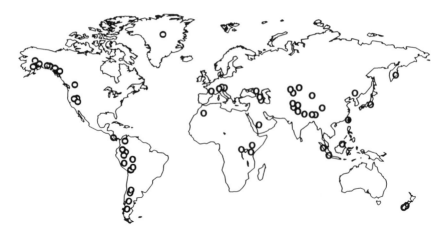

Figure 15.3   Locations of nature reserves with over 3000 m of altitudinal range.

Table 15.3   Potential impacts on mountain nature reserves (over 3000 m) of altitudinal range)

UKMO	GFDL	GISS	OSU
85.5%	64.4%	65.7%	51.3%

Changes in major Holdridge Life Zone (1967) Categories were considered to constitute a potential ecological impact in this analysis.

Figure 15.4   Mountain vegetation regions of the world.

Table 15.4 Potential change in mountain vegetation zones

	UKMO	GFDL	GISS	OSU
Mountain Regions	84.7%	73.9%	64.5%	55.7%
Non-Mountain Regions	71.8%	60.8%	56.1%	49.4%

to the position of the mountain regions with respect to the climatic forcing, or alternatively, can be used as an argument to suggest that the climate models (GCMs) have a higher range of variation in representing the complex climate features of mountainous areas (Halpin, 1993).

Regional methodology

While global-scale GIS analysis of the potential ecological impacts of climatic change offers interesting insights into the distribution of impacts on entire systems of mountain reserve areas, high-resolution, regional case studies must be conducted to develop an understanding of potential changes within individual mountain reserves. A case study of the potential regional impacts of climate change in Costa Rica illustrates some key features of this type of analysis.

The spatial resolution of the global climate data sets outlined above is 0.5° × 0.5° or approximately 55 km × 55 km pixels at the equator in a raster-based GIS system. This spatial resolution is sufficient to represent broad continental features of ecoclimatic transitions, but is entirely inappropriate for use at a regional scale, where questions concerning land management areas are being considered (Halpin, 1992). For example, the entire country of Costa Rica is represented by only 20 pixels in the global ecoclimatic database. To conduct a more meaningful case study in this highly mountainous area, a regional database was developed, with a pixel resolution of approximately 400 m × 400 m, of climate, topography, soils, potential vegetation, vegetation cover, and land use. This data set was better able to sufficiently represent the complex terrain and vegetation features of the region at a 1:500 000 scale using 500 m contours. A second higher-resolution analysis was conducted by collaborators at the Tropical Science Center in Costa Rica, through expert interpretation of climatic and topographic features at a 1:200 000 scale using 100 m contour intervals. This manually-interpreted higher-resolution analysis was digitized into a vector-based GIS for later analysis (Tosi *et al.*, 1992; Secrett, 1992).

Even though Costa Rica covers a relatively small area (51 000 km^2), it is mostly very mountainous, with elevations ranging from sea level to over 3800 m in less than 100 km in the south. The distribution of complex climatic features required the division of the country into five topographically-distinct climatic regions for the development of subregional climatic change models.

Separate lapse rates, sea-level temperatures, and precipitation regimes were interpolated from climate station data for each region. This base climate model was then modified to create two regional climate change sensitivity scenarios (Halpin *et al.*, 1991; Kelly, 1991). A moderate change scenario based on an increase of +2.5 °C temperature and +10% precipitation and a more extreme regional scenario depicting an increase of +3.5 °C and +10% precipitation were used to assess potential changes in tropical montane forest climate zones.

Plate 13 depicts changes in the distribution of ecoclimatic zones for the country as a whole. The base climate mapping of life zones for Costa Rica has been related to specific vegetation patterns through substantial field work over the last 25 years (Tosi, 1969; Holdridge *et al.*, 1971; Sawyer and Lindsey, 1971). A significant amount of change in ecoclimatic zonation occurs under both scenarios for the country, with a 38% change in zones for the 2.5 °C scenario and a 47% change under the 3.6 °C scenario (Halpin *et al.*, 1991). Even larger areas of spatial change in ecoclimatic zones (43% and 60% respectively) were found through interpretation of the higher resolution map by local experts (Tosi *et al.*, 1992).

Potential climatic change expresses only one factor of the two-part global change dilemma outlined in the introduction (Peters and Darling, 1985). Changes in landscape fragmentation are the most noticeable and potent threat to mountain habitats in the humid tropical regions of the world at present (Myers, 1980; Sader and Joyce, 1988; Lugo, 1988). Figure 15.5 illustrates changes in forest cover for Costa Rica derived from field surveys, aerial photography and satellite remote sensing (LANDSAT Thematic Mapper) from 1940–87. Areas of ecoclimatic zone change in non-agricultural regions (60–100% natural vegetation cover), nature reserves, and other special management areas were isolated in the analysis using standard GIS overlay techniques.

Under the 3.5 °C scenario, an ecoclimatic threshold is reached where the distribution of warmer premontane climate zones expand up-slope, lowering the total number of distinct ecoclimatic habitats from nine to six in the southern mountainous region of La Amistad Biosphere Reserve. This potential loss of distinct ecoclimatic zones is depicted in a cross-sectional view in Figure 15.6. The reason for this non-linear change is actually quite simple to explain. The range, or extent of climate now associated with the lowland premontane forests is larger in terms of possible temperature and precipitation combinations than the cooler, narrower vegetation zones at higher altitudes. However, this potential altitudinal extent is not so obvious when the vegetation associated with these warmer ecoclimatic areas is distributed across the low foothill areas of the country. When the sea-level temperatures of the tropical mountains are increased past a threshold temperature range, the distribution of the lowland forest climates can be extended up a larger altitudinal gradient than the significantly narrower ranges of the cooler montane and alpine zones at higher altitudes (Halpin, 1992, 1993). The result

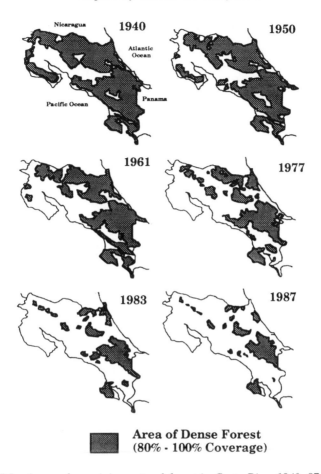

Figure 15.5   Areas of remaining natural forest in Costa Rica, 1940–87.

is that, instead of simply shifting all of the altitudinal zones presently found on the mountain up one evenly-spaced level, the size and composition of the resulting zones change dramatically.

This result is in direct contrast to the present hypothesis that climatic habitats would respond in a regularly ordered pattern. To emphasize this point, the only time one could possibly expect changes in ecoclimatic zones to respond in a regular pattern would be if the climatic limits for each potential vegetation category were identical in their range of temperatures and precipitation responses. All generally accepted vegetation correlation models are based on delineation of differential climate spaces for different types of vegetation categories (Holdridge, 1967; Budyko, 1974; Whittaker, 1975; Box, 1981; Woodward, 1987; Stephenson, 1990; Prentice *et al.*, 1992). Each of these different approaches would predict very different patterns of mountain zonation responses to changing climates due to different interpretations of

Altitudinal Distribution of Eco-climatic Zones

Costa Rica: East-West Transect

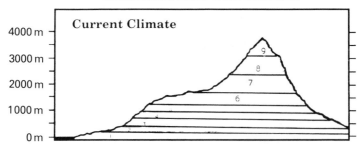

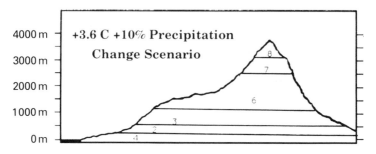

Figure 15.6 A cross-sectional view of changes in ecoclimatic zones along altitudinal gradients in southern Costa Rica under a + 3.6 °C climate change scenario.
1. Premontane Wet Forest: Warm Transition; 2. Tropical Moist Forest; 3. Tropical Wet Forest; 4. Tropical Wet forest: Cool Transition; 5. Premontane Wet Forest: Warm Transition; 6. Premontane Rain Forest; 7. Lower Montane Rain Forest; 8. Montane Rain Forest; 9. Sub-Alpine Rain Paramo.

current vegetation climate ranges for each model. The only common feature that can be generally acknowledged between all models is that none of them will predict climatic zones moving up-slope in a symmetrical, staircase pattern as is now commonly presumed.

Paleo-climatic reconstructions of tropical vegetation from fossil pollen records can be viewed as qualitative climatic change analogies (Flenley, 1979; Vuilleumier and Monasterio, 1986; Peteet, 1987). Figure 15.7 represents a paleovegetation reconstruction of a mountain in Colombia approximately 5° south of the Costa Rica case (from van der Hammen, 1974, and Flenley, 1979). This paleological record qualitatively illustrates the trend from a more narrowly compressed ecoclimatic zonation during a cooler epoch (14 000– 20 000 bp) to broader zonations during a 3° to 6 °C warmer (current climate) period. Similar spatial trends can be seen in other paleoclimate reconstructions

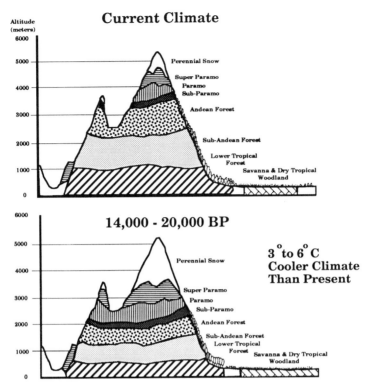

Figure 15.7 A paleoecological reconstruction of vegetation zones of a tropical mountain site in Colombia from a 3–6 °C cooler epoch (14 000–20 000 bp) to warmer, present climate conditions (adapted from Flenley, 1979).

in the neotropical region (Flenley, 1979; Vuilleumier and Monasterio, 1986). This proposed differential expansion of ecoclimatic zonations along altitudinal gradients during different climatic epochs offers an intriguing problem for more rigorous GIS analysis.

Comparative methodology

One of the first geographic questions to ask concerning potential differences in the response of mountain regions to changing climates is: what are the differences in sensitivity of mountains at different latitudes to such change? In order to investigate the sensitivity of different mountain sites to climatic change based on their latitudinal position, three sites ranging from tropical to arctic latitudes were selected for a hypothetical analysis of ecoclimatic zonation changes (Figure 15.8). A 3900 m hypothetical mountain with 100 m elevation intervals was digitized into a raster GIS and used to represent a typical mountain at each site. A single +3.5° temperature and +10 per cent

P.N. Halpin

Boreal/Arctic Mountain Site
Central Alaska (U.S.)
64.00 N / 147.00 W

Temperate Mountain Site
Sierra Nevada (U.S.)
38.00 N / 119 W

Tropical Mountain Site
Central Cordillera (Costa Rica)
9.00 N / 84.00 W

Figure 15.8   Sites for hypothetical mountain climatic zonation change analysis.

precipitation climate change sensitivity scenario was imposed for all sites. This process was done to hold both the topographic effect and magnitude of climatic forcing equal for the GIS experiment. Actual climate data from each site representing La Amistad Biosphere Reserve in Costa Rica, the Sequoia-Kings Canyon Biosphere Reserve in California (USA), and the Denali Biosphere Reserve in Alaska (USA) were used to establish the baseline climates for each hypothetical mountain (Halpin, 1993).

The features of the wet tropical site change have already been outlined above. A loss of climatic zones occurred under the warming scenario, and large bands of lower-elevation tropical premontane forest climatic zones shifted upslope. The cool subalpine paramo climate zone was lost off the top of the mountain under this sensitivity test (Figure 15.9).

The dry temperate mountain site produced a significantly different result, with a loss of two representative climate zones and the expansion of low- and mid-altitude climate ranges. While reduced in coverage, the high-elevation nival (alpine tundra) climate zone at the peak of this temperate site was not lost under this scenario as would be expected under the accepted change paradigm (Figure 15.10).

The cold arctic site begins with a very shallow range of subalpine and alpine forest and tundra sites which expand only slightly up-slope under the same climate scenario. In contrast to the conventional paradigm, the only loss

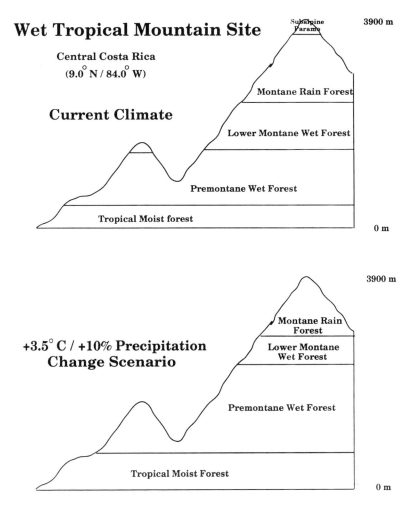

Wet Tropical Mountain Site

Central Costa Rica

(9.0° N / 84.0° W)

Current Climate

Subalpine Paramo

3900 m

Montane Rain Forest

Lower Montane Wet Forest

Premontane Wet Forest

Tropical Moist forest

0 m

3900 m

+3.5° C / +10% Precipitation Change Scenario

Montane Rain Forest

Lower Montane Wet Forest

Premontane Wet Forest

Tropical Moist Forest

0 m

Figure 15.9 Current and changed ecoclimatic zonation for a hypothetical tropical climate mountain located in La Amistad Biosphere reserve, central Costa Rica.

in distinct ecoclimatic zones at the arctic site occurs near the base of the mountain, instead of at the summit (Figure 15.11).

Figure 15.12 illustrates the areas of ecoclimatic change for the three latitudinal sites under the same scenario. This illustration does not represent a new generic paradigm for the expected changes in mountain zones under changing climatic conditions, but instead is presented to caution that changes in ecoclimatic zonations on altitudinal gradients cannot be explained by the global application of simple linear assumptions. The use of extremely straightforward climatic correlation models, in a spatially-explicit GIS modelling framework, leads to significantly more complex solutions than would

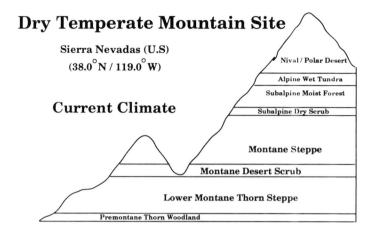

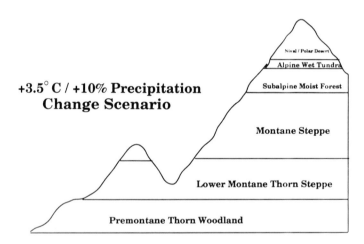

Figure 15.10   Current and changed ecoclimatic zonation for a hypothetical temperate climate mountain located in Sequoia-King Canyon Biosphere Reserve, central California, USA.

generally be anticipated. A new conceptual model based on assumptions of more complex reconfigurations of non-symmetrical ecoclimatic zones should replace current assumptions of symmetrical zonation change for mountain environments under changing environments (Halpin, 1992, 1993).

Conclusions

Any actual impacts of changing climatic conditions on mountain nature reserves will result from a highly complex cascade of environmental and

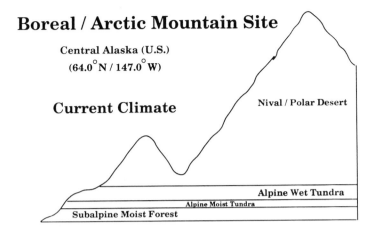

Boreal / Arctic Mountain Site

Central Alaska (U.S.)
$(64.0°N / 147.0°W)$

Current Climate

Nival / Polar Desert

Alpine Wet Tundra

Alpine Moist Tundra

Subalpine Moist Forest

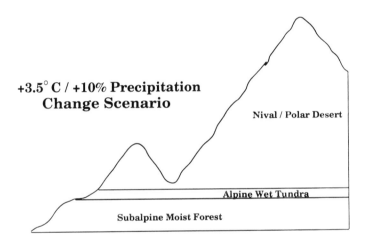

$+3.5°C / +10\%$ **Precipitation Change Scenario**

Nival / Polar Desert

Alpine Wet Tundra

Subalpine Moist Forest

Figure 15.11   Current and changed ecoclimatic zonation for a hypothetical boreal/arctic climate mountain located in the Denali Biosphere Reserve, central Alaska, USA.

ecological feedbacks. In the future, these must be modelled using simulation models that are more physiologically mechanistic and temporally dynamic (Halpin, 1993; Shugart *et al.*, 1992). While significantly more complex to develop, future models of mountain ecosystem dynamics will have to interactively derive environmental site features and changing vegetation states with spatial GIS databases, in order to produce observable ecosystem features which can be validated through field observation or remote sensing. It is to be hoped that further interaction between GIS analysis techniques and ecological theory development will hopefully help to sharpen our understanding of the complex ecological processes which control mountain environments.

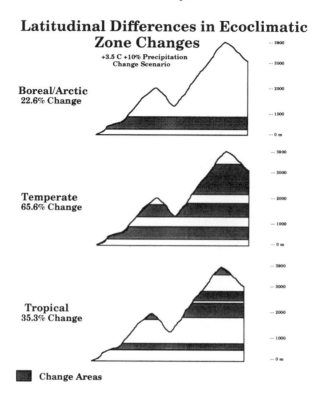

Figure 15.12   Areas of ecoclimatic change for hypothetical arctic, temperate and tropical mountains.

References

Box, E.O., 1981, *Macroclimate and Plant Forms: An Introduction to Predictive Modeling in Phytogeography*, Hague: Junk.

Budyko, M.I., 1974, *Climate and Life*. International Geophysics Series, Vol. 18. New York: Academic Press.

Cramer, W. and Prentice, I.C., 1992, Simulation of regional soil moisture deficits on a European scale, *Norsk Geografisk Tidsskrift*.

Davis, M.B., 1989, Lags in vegetation response to greenhouse warming, *Climatic Change*, **15**, 75–82.

Emanuel, W.R., Shugart, H.H. and Stevenson, M.P., 1985, Climatic change and the broad-scale distribution of terrestrial ecosystem complexes, *Climatic Change*, **7**, 29–43.

Flenley, J.R., 1979, *The Equatorial Rain Forest: A Geological History*, London: Butterworths.

Graham, R.W., 1988, The role of climate change in the design of biological reserves: The paleoecological perspective for conservation biology, *Conservation Biology*, **2**, 391–4.

Halpin, P.N., 1992, Potential impacts of climate change on protected areas: global assessments and regional analysis, in IUCN, *Proceedings of the IV World Parks Congress*, Caracas, Venezuela, in press.

Halpin, P.N., 1994, Latitudinal variation of montane ecosystem response to potential climatic change, in Beniston, M. (Ed.) *Mountain Environments in Changing Climates*, London: Routledge, in press.

Halpin, P.N. and Secrett, C.M., 1992, Potential impacts of climate change on forest protection in the humid tropics: A case study in Costa Rica, *Proceedings of the Symposium on Impacts of Climate Change on Ecosystems and Species*, Amersfoort, Netherlands (in press).

Halpin, P.N., Kelly, P.M., Secrett, C.M. and Smith, T.M., 1991, Climate Change and Central American Forest Systems: Costa Rica Pilot Project (symposium report).

Hansen, J., Fung, I., Lacis, A., Lebedef, S., Rind, D., Ruedy, R., Russel, G. and Stone, P., 1988, Global climate changes as forecast by the Goddard Institute for Space Studies three dimensional model, *Journal of Geophysical Research*, **93**, 9341–64.

Holdridge, L.R., 1967, *Life Zone Ecology*, San Jose, Costa Rica: Tropical Science Center.

Holdridge, L.R., Grenke, W.C., Hatheway, W.H., Liang, T. and Tosi, J.A. Jr., 1971, *Forest Environments in Tropical Zones: A Pilot Study*, Oxford: Pergamon Press.

Hudson, W.E., 1991, *Landscape linkages and biodiversity*, Washington, DC: Island Press.

Hunter, M.L., Jacobson, G.L. Jr. and Webb, T. III., 1988, Paleoecology and the Coarse Filter Approach to Maintaining Biological Diversity, *Conservation Biology*, **2**, 375–85.

IPCC, 1989, 'Unmanaged ecosystems – biological diversity, adaptive responses to climate change,' unpublished Working Paper, Resource Use and Management Subgroup of IPCC Working Group III.

Kalkstein, L.S., 1991, *Global comparisons of selected GCM control runs and observed climate data*, Washington, DC: US Environmental Protection Agency.

Kelly, P.M., 1991, 'Regional Climate Change Scenarios for Costa Rica,' unpublished project report, Climatic Research Unit, University of East Anglia.

Leemans, R. and Cramer, W.P., 1990, *The IIASA database for mean monthly values of temperature, precipitation and cloudiness on a global terrestrial grid*, Report WP-90-41, Laxenburg: International Institute for Applied Systems Analysis.

Leemans, R. and Halpin, P.N., 1992, Biodiversity and global change, in Groombridge, B. (Ed.) *Biodiversity Status of the Earth's Living Resources*, London: Chapman and Hall and World Conservation Monitoring Center.

Lugo, A.E., 1988, Estimating reductions in the diversity of tropical forest species, in Wilson, E.O. and Peter, F.M. (Eds) *Biodiversity*, Washington, DC: National Academy Press, 58–70.

MacArthur, R.H., 1972, *Geographical Ecology*, New York: Harper and Row.

Mackintosh, G., Fitzgerald, J. and Kloepfer, D., 1989, *Preserving Communities and Corridors*, Washington, DC: Defenders of Wildlife and G.W. Press.

Manabe, S. and Wetherald, R.T., 1987, Large scale changes in soil wetness induced by an increase in carbon dioxide, *Journal of Atmospheric Science*, **44**, 1211–35.

McNeely, J.A., 1990, Climate change and biological diversity: policy implications, in Boer, M.M. and de Groot, R.S. (Eds), *Landscape – Ecological Impact of Climate Change*, Amsterdam: IOS Press.

Mitchell, J.F.B., 1983, The seasonal response of a general circulation model to changes in CO2 and sea temperatures, *Quarterly Journal of the Royal Meteorological Society*, **109**, 113–52.

Myers, N., 1980, The present status and future prospects of tropical moist forests, *Environmental Conservation*, **7**, 101–14.

Noss, R., 1992. The wildlands project: land conservation strategy, in *Wild Earth*, New York: Cenozoic Society.

Peteet, D., 1987, Late Quaternary vegetation and climatic history of the montane and lowland tropics, in Rosenzweig, C. and Dickinson, R. (Eds) *Climate-Vegetation Interactions*, Boulder: University Corporation for Atmospheric Research, 72–6.

Peters, R.L. and Darling, J.D., 1985, The greenhouse effect and nature reserves, *Bioscience*, **35**, 707–17.

Prentice, I.C., Webb, R.S., Ter-Mikhalaelian, M.T., Solomon, A.L., Smith, T.M., Pitovranov, S.E., Nikolov, N.T., Minin, A.A., Leemans, R., Lavorel, S., Korzukhin, M.D., Hrabovszky, J.P., Helmisaari, H.O., Harrison, S.P., Emanuel, W.R. and Bonan, G.B., 1989, Developing a Global Vegetation Dynamics Model: Results of the IIASA Summer Workshop. IIASA, Laxenburg, Austria, 49pp.

Prentice, I.C., Cramer, W., Harrison, S.P., Leemans, R., Monserud, R.A. and Solomon, A.M., 1992, A global biome model based on plant physiology and dominance, soil properties and climate, *Journal of Biogeography*, **19**, 117–34.

Sader, S.A. and Joyce, A.T., 1988, Deforestation rates and trends in Costa Rica: 1940 to 1983, *Biotropica*, **20**, 11–19.

Sawyer, J. and Lindsey, A., 1971, *Vegetation of the Life Zones in Costa Rica*, Indiana Academy of Science Monograph 2.

Schlesinger, M. and Zhao, Z., 1988, *Seasonal climatic changes induced by doubled CO2 as simulated by the OSU atmospheric GCM/mixed layer ocean model*, Corvallis: Climate Research Institute, Oregon State University.

Scott, M.J., Davis, F., Csuti, B., Butterfield, B., Noss, R., Caico, S., Anderson, H., Ulliman, J., D'Erchia, F. and Groves, C., 1990, Gap Analysis: Protecting Biodiversity Using Geographic Information Systems, University of Idaho.

Secrett, C.M., 1992, *Adapting to Climate Change; A Strategy for the Tropical Forest Sector*, Project Paper, London: International Institute for Environment and Development.

Sedjo, R.A. and Solomon, A.M., 1989, Climate and forests, in Rosenberg, N.J., Easterling, W.E., Crosson, P.R. and Darmstadter, J. (Eds) *Greenhouse Warming: Abatement and Adaptation*, Washington, DC: Resources For the Future, 105–19.

Shaffer, C.L., 1990, *Nature Reserves: Island Theory and Conservation Practice*, Washington, DC: Smithsonian Institute.

Shugart, H.H., Smith, T.M. and Post, W.M., 1992, The potential for application of individual-based simulation models for assessing the effects of global change, *Annual Review of Ecology and Systematics*, **23**, 15–38.

Smith J. and Tirpak, D., 1989, *The Potential Effects of Climate Change on the United States*, Report to Congress, Washington, DC: US Environmental Protection Agency.

Smith, T.M., Shugart, H.H. and Halpin, P.N., 1990, Global Forests, in *Progress Reports on International Studies of Climate Change Impacts*, Washington, DC: US Environmental Protection Agency.

Smith, T.M., Shugart, H.H., Bonan, G.B. and Smith, J.B., 1992, Modeling the potential response of vegetation to global climate change, *Advances in Ecological Research*, **22**, 93–113.

Solomon, A.M., 1986, Transient responses of forests to CO2-induced climate change: Simulation modeling experiments in eastern North America, *Oecologia*, **68**, 567–9.

Stephenson, N.L., 1990, Climatic control of vegetation distribution: the role of the water balance, *The American Naturalist*, **135**, 649–70.

Tosi, J.A. Jr., 1969, *Mapa Ecologico Republica de Costa Rica*, San Jose, Costa Rica: Tropical Science Center.

Tosi, J.A., Watson, V. and Echeverria, J., 1992, *Potential Impacts of Climate Change on the Productive Capacity of Costa Rican Forests: A Case Study*, San Jose, Costa Rica: Tropical Science Center.

van der Hammen, T., 1974, The Pleistocene changes of vegetation and climate in tropical South America, *Journal of Biogeography*, **1**, 3–26.

Vuilleumeir, F. and Monasterio, M., 1986, *High Altitude Tropical Biogeography*, New York: Oxford University Press.

Whittaker, R.H., 1975, *Communities and Ecosystems*, 2nd edn, New York: Macmillan.
Woodward, F.I., 1987, *Climate and Plant Distribution*, Cambridge: Cambridge University Press.
World Conservation Monitoring Center, 1992, *Global Biodiversity: Status of the Earth's Living Resources*, London: Chapman & Hall.

Index